Sustainability 4.0

This book presents a fresh perspective by exploring the various layers of sustainability within the whole manufacturing and incorporating Industry 4.0 and smart manufacturing principles. The concept can be applied to manufacturing systems, focusing on core, chosen, and given layers. Each layer of the system manufacturing can be analyzed independently to identify key components and evaluate their sustainability impact.

Sustainability 4.0: Objectives, Challenges, and Expectations provides a thorough examination and profound understanding of Sustainability 4.0, a crucial goal set forth by the United Nations (UN). This book introduces an innovative index created to measure sustainability and sustainable development within manufacturing enterprise systems. The index evaluates machine design, manufacturing processes, and manufacturing systems. A distinct methodology is outlined also for assessing an enterprise's sustainability concerning Industry 4.0 including emerging technologies, technology selection, adoption, and site readiness and implementation. This book discusses the impact of emerging technologies associated with Industry 4.0 on sustainability practices, as well as a flexible framework for evaluating the sustainability of operations amidst technological advancements and evolving industry standards.

This book is recommended as a valuable reference for senior-year undergraduate students, graduate students, postgraduate students, academic researchers, and industrial practitioners. The insights offered within its pages are particularly beneficial for technical audiences in search of solutions, case studies, and prototype implementations.

Manufacturing and Production Engineering Series

Series Editors: Hamid R. Parsaei, Texas A&M University, & Waldemar Karwowski, University of Central Florida

This series will provide an outlet for the state-of-the-art topics in manufacturing and production engineering disciplines. This new series will also provide a scientific and practical basis for researchers, practitioners, and students involved in areas within manufacturing and production engineering. Issues envisioned to be addressed in this new series would include, but not limited to, the following: Additive Manufacturing, 3D Visualization, Mass Customization, Material Processes, Cybersecurity, Data Science, Automation and Robotics, Underwater Autonomous Vehicles, Unmanned Autonomous Vehicles, Robotics and Automation, Six Sigma and Total Quality Management, Manufacturing Cost Estimation and Cost Management, Industrial Safety, Programmable Logic Controllers, to name just a few.

Decision Making in Risk Management: Quantifying Intangible Risk Factors in Projects
Christopher Cox

Reconfigurable Manufacturing Enterprises for Industry 4.0
Ibrahim H. Garbie and Hamid R. Parsaei

Recent Advances in Time Series Forecasting
Dinesh C. S. Bisht and Mangey Ram

Cost Analysis for Engineers and Scientists
Fraiborz (Fred) Tayyari

Sustainability 4.0: Objectives, Challenges, and Expectations
Ibrahim H. Garbie and Hamid R. Parsaei

For more information on this series, please visit: https://www.routledge.com/Manufacturing-and-Production-Engineering/book-series/CRCMNPRDENG

Sustainability 4.0
Objectives, Challenges, and Expectations

Ibrahim H. Garbie and Hamid R. Parsaei

CRC Press
Taylor & Francis Group
Boca Raton London New York

CRC Press is an imprint of the
Taylor & Francis Group, an **informa** business

Designed cover image: Shutterstock

First edition published 2026
by CRC Press
2385 NW Executive Center Drive, Suite 320, Boca Raton FL 33431

and by CRC Press
4 Park Square, Milton Park, Abingdon, Oxon, OX14 4RN

CRC Press is an imprint of Taylor & Francis Group, LLC

ISBN: 978-1-041-08198-2 (hbk)
ISBN: 978-1-041-08319-1 (pbk)
ISBN: 978-1-003-64483-5 (ebk)

DOI: 10.1201/9781003644835

Typeset in Times
by codeMantra

Contents

Preface x
About the Authors xii

PART 1 Introduction

Chapter 1 Introduction to Sustainability 4.0 3

1.1 Sustainable Development Goals (SDGs) 3
1.2 Sustainable Manufacturing (SM4.0) 5
1.3 Industry 4.0 (I4.0) 6
1.4 Linking between SM4.0 and SI4.0 7
1.5 Sustainability 4.0 Perspectives 9
1.6 Outline and Scope of this Book 11
1.7 Concluding Remarks 11
References 12

PART 2 Sustainability 4.0 Perspectives

Chapter 2 Sustainability 4.0 from a Technical Perspective 17

2.1 Architectural Perspective of Sustainable Manufacturing 17
2.1.1 Introduction of Sustainable Manufacturing 17
2.1.2 Core Layer: Machine Design (MD) 17
2.1.3 Chosen Layer: Manufacturing Processes (MP) 17
2.1.4 Given Layer: Manufacturing Systems (MS) 18
2.2 Architectural Perspective of Industry 4.0 18
2.2.1 Introduction of Industry 4.0 18
2.2.2 Core Layer: Virtual Environment (VE) 19
2.2.3 Chosen Layer: Technologies I4.0 Selection (TS) 20
2.2.4 Given Layer: Site Readiness (RS) 20
2.3 Concluding Remarks 20
References 20

Chapter 3 Sustainability 4.0 from a Human Perspective 22

3.1 Sustainable Manufacturing 22
3.2 Industry 4.0 25
3.3 Numerical Example 3.1 27
3.4 Concluding Remarks 30
References 30

PART 3 Sustainable Manufacturing

Chapter 4 Architecture System of Sustainable Manufacturing 33

4.1 Introduction and Background 33
4.2 Manufacturing Sustainability Layers 34
4.2.1 Machine Component Design (Core Layer) 35
4.2.2 Manufacturing Processes (Chosen Layer) 35
4.2.3 Manufacturing Systems (Given Layer) 35
4.3 The Triple Bottom Line of Manufacturing Sustainability 36
4.4 Concluding Remarks 37
References 38

Chapter 5 Sustainable Machine/Product Design 39

5.1 Introduction to Machine Design (MD) 39
5.2 Machine/Product Design within an Architecture Perspective 40
5.2.1 Sustainable Machine Design Challenges and Solutions 40
5.3 Analysis and Assessment of Sustainable Machine/Product Design (MD) 41
5.4 Numerical Example 5.1 47
5.5 Concluding Remarks 48
References 50

Chapter 6 Sustainable Manufacturing Processes 51

6.1 Introduction to Manufacturing Processes (MP) 51
6.2 Manufacturing Processes within an Architecture Perspective 52
6.3 Analysis and Assessment of Sustainable Manufacturing Processes 53
6.4 Numerical Example 6.1 56
6.5 Concluding Remarks 57
References 58

Chapter 7 Sustainable Manufacturing Systems 59

7.1 Introduction to Manufacturing Systems (MS) 59
7.2 Manufacturing Systems within an Architecture Perspective 60
7.3 Analysis and Assessment of Sustainable Manufacturing Systems 61
7.4 Numerical Example 7.1 63
7.5 Concluding Marks 65
References 65

Chapter 8 Manufacturing Sustainability 67

8.1 Introduction to Manufacturing Sustainability 67
8.2 Assessment of Manufacturing Sustainability 68
8.3 Numerical Example 8.1 68
8.4 Concluding Remarks 72
References 73

PART 4 Sustainable Industry 4.0

Chapter 9 Sustainable Emerging Technologies 77

9.1 Introduction to Emerging Technologies 77
9.2 Components of Emerging Technologies 78
9.2.1 Digitalization or Digital Manufacturing (DI) 79
9.2.2 IoT or Industrial Internet of Things (IIoT) 79
9.2.3 Big Data Analytics 80
9.2.4 CC or CM 80
9.2.5 CS and Data Privacy 80
9.2.6 VR/AR 81
9.2.7 Additive Manufacturing (AM) 81
9.2.8 Cyber-Physical Production Systems 81
9.3 Analysis and Assessment of Sustainable Emerging Technologies (ET) 82
9.4 Numerical Example 9.1 84
9.5 Concluding Remarks 84
References 86

Chapter 10 Sustainable Technologies Selection 87

10.1 Introduction to Selection Technologies 87
10.2 Design Principles of Industry 4.0 88
10.2.1 Interoperability (IN) 88
10.2.2 Virtualization (Virtual Reality) (VI) 88
10.2.3 Decentralization (DE) 89
10.2.4 Real-Time Capability (RT) 89
10.2.5 Service Orientation (SO) 89
10.2.6 Modularity (MO) 89
10.3 Mapping between Emerging Technologies and Design Principles 90
10.3.1 Emerging Technologies Categorization 90
10.3.2 Mapping Emerging Technologies to Design Principles 90
10.4 Analysis and Assessment of Sustainable Technologies 4.0 Selection (TS) 94

10.5 Numerical Example 10.1 94
10.6 Concluding Remarks 97
References 98

Chapter 11 Sustainable Site Readiness/Implementation 100

11.1 Introduction to Site Readiness/Implementation 100
11.2 Issues of Site Readiness 100
11.2.1 Technical Challenges of Implementation 102
11.2.2 Risks 103
11.2.3 Critical Success Factors 104
11.2.4 Opportunities 106
11.2.5 Enablers of Implementation 107
11.3 Analysis and Assessment of Sustainable Site Readiness 108
11.3.1 For Technical Challenges of Implementation (h_1) 110
11.3.2 For Risks (h_2) 111
11.3.3 For Critical Success Factors (h_3) 111
11.3.4 For Opportunities (h_4) 111
11.3.5 For Enablers of Implementation (h_5) 111
11.4 Numerical Example 11.1 112
11.5 Concluding Remarks 116
References 116

Chapter 12 Industry 4.0 Sustainability 118

12.1 Introduction 118
12.2 Industry 4.0/Smart Manufacturing Streams 119
12.2.1 Emerging Technologies (ET) Sustainability 119
12.2.2 Technologies Selection (TS) Sustainability 120
12.2.3 Site Readiness (SR) Sustainability 121
12.3 Sustainability Assessment of Industry 4.0 121
12.4 Numerical Example 12.1 122
12.4.1 For the Emerging Technologies (ET) 122
12.4.2 For the Technologies Selection (TS) 122
12.4.3 For the Site Readiness (SR) 124
12.5 Concluding Remarks 126
References 126

PART 5 Sustainability 4.0

Chapter 13 Effects of Industry 4.0 Technologies on Sustainability 129

13.1 Introduction 129
13.2 Sustainability Pillars of Manufacturing Industries 130

13.2.1 Pillar Integration and Features 130
13.2.2 Short-Term Features 130
13.2.3 Long-Term Features 131
13.2.4 Pillar Issues, Indicators, and Metrics 131
13.3 Industry 4.0/Smart Manufacturing 132
13.3.1 Industry 4.0 Technologies Overview 132
13.3.2 Adoption beyond Automation 133
13.3.3 Digitization and Big Data 135
13.3.4 Impact on Sustainability 135
13.4 Concluding Remarks 137
13.4.1 Economic Sustainability 137
13.4.2 Social Sustainability 137
13.4.3 Environmental Sustainability 138
References 138

Chapter 14 Building Sustainability 4.0 140

14.1 Introduction 140
14.2 Linking the Pillars of Core Layer of Sustainability 4.0 141
14.3 Perspective Building of Sustainability 4.0 142
14.4 Numerical Example #1 144
14.4.1 Linking the Pillars of A, P, and G 144
14.4.2 Industry 4.0 Sustainability 146
14.4.3 Manufacturing Sustainability 149
14.5 Concluding Remarks 153
References 154

Index 155

Preface

This book summarizes the author's research on sustainability and Industry 4.0 after publishing two books related to sustainability in manufacturing and reconfiguration of manufacturing enterprises for achieving Industry 4.0. This third book aims to integrate sustainability and Industry 4.0 and its corresponding smart manufacturing to participate in or create a new concept called Sustainability 4.0 (S4.0). The new title is *Sustainability 4.0: Objectives, Challenges, and Expectations*, which refers to topics from sustainability in manufacturing (SM4.0) and sustainability in adopting and implementing Industry 4.0/smart manufacturing (I4.0).

The value addition made in this book is the integration of sustainable manufacturing with Industry 4.0. This book introduces a novel perspective of manufacturing industries is so-called "Sustainability 4.0" by examining the sustainability/sustainable development (S/SD) and Industry 4.0 of connecting manufacturing layers and architecture within the context of sustainable manufacturing and Industry 4.0. It presents the concepts of layers and architecture through the core, chosen, and given layers, and each layer is dissected, identifying essential components and assessing their sustainable index independently. The manufacturing layers emerge as a pivotal force, influencing individual layers and decision-making and shaping the final layers.

This book investigates and analyzes the theoretical approaches and techniques for presenting the relationships between sustainability/sustainable development issues and requirements of Industry 4.0 through a novel perspective of Sustainability 4.0. The title describes Sustainability 4.0 in manufacturing industries from a technical perspective and a human perspective. Layers and architecture of sustainable manufacturing will be discussed through three main layers: design of machine and product, analysis and investigation of manufacturing processes, and a full perspective view of designing manufacturing systems. Also, layers of implementing and/or adopting Industry 4.0 will be presented through emerging technologies, either advanced manufacturing technologies or advanced digital technologies for Industry 4.0, technological adoption and selection, and site readiness and implementation. The indexes and values are estimated through Sustainability 4.0 objectives, challenges, and expectations toward the requirements of manufacturing sustainability and Industry 4.0 implementation.

This book is logically organized and succinct based on sustainable manufacturing (SM4.0) and Industry 4.0 (I4.0). Based on the above description, this book has 14 chapters, which are classified into five parts. Part 1 includes Chapter 1, focusing on the introduction and background of sustainability/sustainable development, Industry 4.0, and Sustainability 4.0 perspectives. Part 2 consists of two chapters on Sustainability 4.0 in manufacturing in terms of technical perspective, as in Chapter 2, and human perspective, as in Chapter 3. Based on this, Part 3 contains five chapters that mention the sustainability issues regarding the whole manufacturing taking into consideration the Sustainability 4.0 approach. Chapter 4 presents the general concepts of the whole sustainable manufacturing. Sustainability in the whole of manufacturing industries (machine design, analysis of manufacturing processes, and facilities design and analysis) will be introduced in Chapters 5–7. Chapter 8 proposes a manufacturing sustainability assessment in the whole manufacturing sector.

In Part 4, the sustainability of adoption of the emerging technologies, technologies selection, and implementation of Industry 4.0 are explained through four chapters. Chapter 9 discusses emerging technologies to achieve greater efficiency in manufacturing operations for creating the virtual environment. Chapter 10 explains how to adopt and select emerging technologies. Site readiness of Industry 4.0 and implementation will be presented in Chapter 11. Chapter 12 proposes the Industry 4.0 sustainability assessment. Part 5 highlights incorporating Industry 4.0 with sustainability in manufacturing through two chapters. Chapter 13 focuses on integrating and effecting Industry 4.0 on manufacturing sustainability, while Chapter 14 focuses on building Sustainability 4.0.

Also, most of the chapters are backed by background literature. This book is suitable reading for senior-year undergraduate students, graduate students, postgraduate students, academic researchers, and industrial practitioners as a source of information. Finally, the insights of this book could be valuable for the technical audience as a solution, case study, and prototype for implementation.

Ibrahim H. Garbie
Hamid R. Parsaei

About the Authors

Ibrahim H. Garbie is a Professor of Industrial Engineering at Helwan University, Egypt, and he is well recognized as a leader in advanced manufacturing systems. He has authored over 100 articles in well-regarded international peer-reviewed archival journals, conference proceedings, technical reports, and book chapters in manufacturing systems design, manufacturing strategies, reconfiguration, sustainability, and Industry 4.0. Furthermore, he serves on the editorial board of several journals and has been a reviewer of over two dozen international peer-reviewed journals. His second book (co-authored with Hamid R. Parsaei), *Reconfigurable Manufacturing Enterprises for Industry 4.0* (CRC Press, 2022), received the 2022 IISE Joint Publishers Book-of-the-Year Award.

Hamid R. Parsaei, PE, is a Professor of Industrial and Systems Engineering and the Director of Accreditation and Assessment for the College of Engineering at Texas A&M University. He also served as Associate Dean of Academic Affairs at the Texas A&M University Qatar campus. Before joining Texas A&M University, he served as a Professor and Chair of the Department of Industrial Engineering at the University of Houston for 10 years. He is a registered professional engineer in Texas and a Life Fellow of the Institute of Industrial and Systems Engineers (IISE), American Society for Engineering Education (ASEE), Society of Manufacturing Engineers (SME), and Industrial Engineering and Operations Management (IEOM) Society International. Additionally, he served as the President of the Industrial Engineering and Operations Management (IEOM) Society International from 2021 to 2023. His primary research interests include advanced manufacturing systems, Industry 4.0, and economic and financial decision-making. He is an active ABET Program Evaluator (PEV) since 2004 and served as a commissioner of the Engineering Accreditation Commission of ABET. He is an ABET Member of the Board of Delegates representing the Institute of Industrial and Systems Engineering. He has published over 350 articles in peer-reviewed archival journals and conference proceedings and has served as principal or co-principal investigator for projects funded by the federal government and private sector. His most recent book, *Reconfigurable Manufacturing Enterprises for Industry 4.0* (CRC Press, 2022) (jointly authored by Ibrahim H. Garbie), received the 2022 IISE Joint Publishers Book-of-the-Year Award.

Part 1

Introduction

1 Introduction to Sustainability 4.0

1.1 SUSTAINABLE DEVELOPMENT GOALS (SDGs)

In September 2015, the United Nations General Assembly identified 17 sustainable development goals (SDGs) as part of its Agenda for Action. These goals serve as a global framework to promote prosperity while protecting the planet. By balancing the economic, social, and environmental dimensions of sustainability—often referred to as the "triple bottom line" (TBL)—the SDGs aim to transform the world by 2030.

The 17 SDGs and their key objectives are as follows:

Goal 1: No Poverty: end poverty in all its forms everywhere.

Goal 2: Zero Hunger: achieve food security and improved nutrition, and promote sustainable agriculture.

Goal 3: Good Health and Well-Being: ensure healthy lives and promote well-being for all at all ages.

Goal 4: Quality Education: ensure inclusive and equitable quality education and promote lifelong learning opportunities for all.

Goal 5: Gender Equality: achieve gender equality and empower all women and girls.

Goal 6: Clean Water and Sanitation: ensure availability and sustainable management of water and sanitation for all.

Goal 7: Affordable and Clean Energy: ensure access to affordable, reliable, sustainable, and modern energy for all.

Goal 8: Decent Work and Economic Growth: promote sustained, inclusive, and sustainable economic growth, full and productive employment, and decent work for all.

Goal 9: Industry, Innovation, and Infrastructure: build resilient infrastructure, promote inclusive and sustainable industrialization, and foster innovation.

Goal 10: Reduced Inequality: reduce inequality within and among countries.

Goal 11: Sustainable Cities and Communities: make cities and human settlements inclusive, safe, resilient, and sustainable.

Goal 12: Responsible Consumption and Production: ensure sustainable consumption and production patterns.

Goal 13: Climate Action: take urgent action to combat climate change and its impacts.

Goal 14: Life below Water: conserve and sustainably use the oceans, seas, and marine resources for sustainable development.

DOI: 10.1201/9781003644835-2

Goal 15: Life on Land: protect, restore, and promote sustainable use of terrestrial ecosystems, sustainably manage forests, combat desertification, halt and reverse land degradation, and halt biodiversity loss.

Goal 16: Peace, Justice, and Strong Institutions: promote peaceful and inclusive societies for sustainable development, provide access to justice for all, and build effective, accountable, and inclusive institutions at all levels.

Goal 17: Partnerships for the Goals: strengthen the means of implementation and revitalize the global partnership for sustainable development" (UN, 2015).

While all 17 goals are interconnected, some have a direct bearing on industrial sustainability, which is the focus of this book. Specifically, SDGs 7, 8, 9, 12, and 17 are most relevant to Sustainable Manufacturing (SM4.0). These goals emphasize energy efficiency, economic growth, innovation, sustainable production practices, and collaborative partnerships.

Other SDGs, such as SDGs 1, 2, 3, 4, 5, 10, 11, and 16, address broader social objectives, while SDGs 6, 13, 14, and 15 focus on biological and environmental challenges. Together, these goals create a comprehensive framework for sustainable development that will shape industries, policies, and practices worldwide (Pansare et al., 2023) (Figure 1.1). By 2030, achieving these goals will require collective efforts from governments, industries, academia, and civil society. This book explores how engineering professionals can contribute to these global objectives through the lens of Sustainability 4.0.

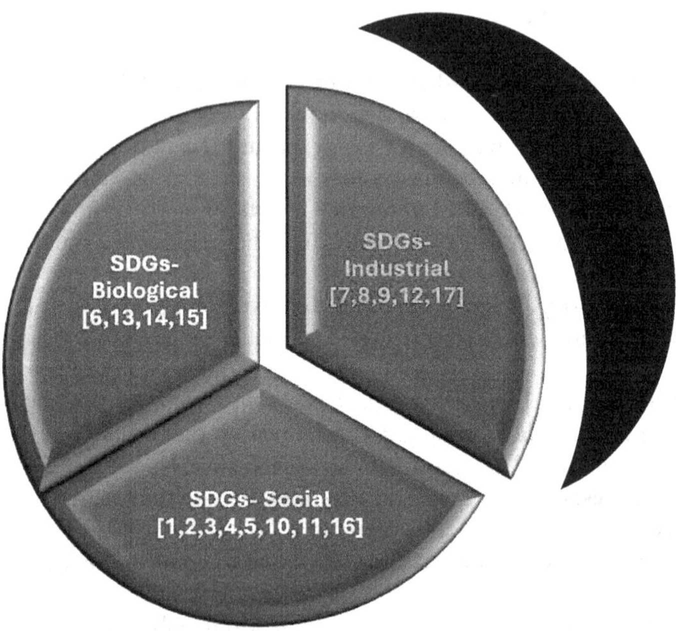

FIGURE 1.1 The sustainable development goals (SDGs) divisions.

1.2 SUSTAINABLE MANUFACTURING (SM4.0)

Manufacturing industries play a vital role in growing the economy of any country, either developed or emerging or developing. Therefore, it is essential to explore the critical factors or parameters that lead to manufacturing sustainability in their companies (Yadav et al., 2020). Sustainable manufacturing (SM4.0) is responsible for achieving SDGs 7, 9, and 12 from the list given in Table 1.1. Although there is no universal definition for sustainability in manufacturing industries, it was defined as green manufacturing products, processes, and systems; ecological balance as the environmental management; and optimizing in consumption of natural resources (Garbie, 2013, 2014). Manufacturing still has a strong base of urgent importance to encourage and motivate other sectors of the national economy to develop and update their activities and to inspire research and development (Garbie, 2013, 2014). The developed countries (USA, UK, Germany, Italy, etc.) are more adopting manufacturing sustainability in their industries than emerging and developing countries in the rate of sustainability adoption and implementation (Yadav et al., 2020).

There is a debate question about the relationship between sustainability and engineering. This relationship exists as engineering represents a major task of achieving sustainability. Science can also participate in this point of view as it generates true knowledge and facts. Engineering is more significant than science in this matter because not only is it responsible for changing the world but also is responsible for keeping these systems around us sustainable and surviving

TABLE 1.1
Sustainable Development Goals (SDGs)

Goal	Classification
1	Social sustainable development goals
2	
3	
4	
5	
6	Biological sustainable development goals
7	Industrial sustainable development goals
8	
9	
10	Social sustainable development goals
11	
12	Industrial sustainable development goals
13	Biological sustainable development goals
14	
15	
16	Social sustainable development goals
17	Industrial sustainable development goals

(Hasna, 2010; Garbie, 2014). Therefore, it can be said that manufacturing sustainability is an "Engineering Responsibility".

Sustainability/sustainable development of manufacturing industries should consider issues such as expanding by increasing resources (people, equipment, etc.) and lack of skilled employees. Achieving manufacturing sustainability in industries requires a full perspective not just on the designing or product components and the corresponding associated manufacturing processes but also the facilities design and layout representing in types of manufacturing systems (Garbie, 2013, 2014).

Sustainable manufacturing is a new concept that appeared in the last two decades, concerning manufacturing industries in terms of survival. Manufacturing sustainability is one of the most important forms to deal with any domestic and international financial crisis. The manufacturing sustainability models will be discussed by incorporating all aspects of sustainable issues, including the sustainability assessment index. The major goal of this book is to consider manufacturing industries' requirements.

1.3 INDUSTRY 4.0 (I4.0)

Industry 4.0 (I4.0) is considered an innovation in the last decade that is focused on implementing or adopting emerging or contemporary new manufacturing technologies in industries. SDG Number Nine is titled "Build resilient infrastructure, promote inclusive and sustainable industrialization, and foster innovation". Industry 4.0 was initially proposed by Germany in 2011, at the same time in the USA under the title Smart Manufacturing. Therefore, the words "Industry 4.0 and Smart Manufacturing" have overlapping definitions and meanings. The major target of adopting Industry 4.0 is to integrate advanced manufacturing technologies and digital technologies with traditional manufacturing technologies and processes under the umbrella of cyber-physical production systems (CPPS). Industry 4.0 represents a new industrial paradigm. Adopting Industry 4.0 technologies in manufacturing is playing a big role in industrialized countries. Developed countries have been eager to capture Industry 4.0 technologies strategy. Emerging (e.g., BRICS) and developing economies (newly industrialized countries) have been actively embracing and adopting emerging advanced digital and manufacturing I4.0 technologies.

It was noticed that Industry 4.0 activities are incorporated and engaged by most of the manufacturing companies, especially in advanced manufacturing and digital technologies in developed and emerging countries. A little bit of conducted survey or research in this manner was started in the developing countries. Most industrial organizations in developing countries are still working in the "Industry 2.0~3.0" period, and the requirements for adoption and implementing I4.0 have not yet started. This means that developing countries and some emerging countries have not completed the physical infrastructure and require digital technologies. It is noticed that these countries are further from the Industry 4.0 technologies. It seems that in emerging and developing countries, most manufacturing industry companies are not ready for Industry 4.0 adoption and implementation. Therefore, these manufacturing companies hope to use these digital and manufacturing technologies revolutions to leapfrog the traditional stage of the industry from Industry

2.0 and Industry 3.0 into Industry 4.0. Small- and medium-sized manufacturing enterprises in emerging countries (for example, China and India) believe that their demand for I4.0 technologies will be inhibited because of many contributing factors. Management commitment, extensive implementation costs, unemployment, skilled employees, and a lack of organizational knowledge and process changes in management systems represent the barriers and obstacles to implementing Industry 4.0.

These emerging technologies are advanced manufacturing technologies and advanced digital technologies. Advanced manufacturing technologies include additive manufacturing (AM) (e.g., 3D printing), updating robotics systems, and sensors and actuators. Advanced digital technologies consist of cloud computing/cloud manufacturing (CC/CM), big data Analytics (BDA), the Internet of Things (IoT), and blockchain.

The selection of Industry 4.0 technologies and implementation of Industry 4.0 seem to an overlapping definitions and meanings, but they are different. The selection of Industry 4.0 technologies is the dynamic decision to introduce advanced digital or manufacturing technologies feasible with the type of manufacturing industry, but the implementation of Industry 4.0 is the deployment of these advanced digital or manufacturing technologies. In this book, we are trying to explore or focus on these two major points. Adopting Industry 4.0 technologies will create smart plants, either advanced digital or advanced manufacturing. The deployment and implementation of Industry 4.0 technologies have several benefits, including reducing product costs, improving lead times, and product quality (Kamble and Gunasekaran, 2019).

Implementing Industry 4.0 technologies requires full support from government programs (Machado et al., 2020), "

- **Germany**: 'High Tech Strategy 2020' and 'Industry 4.0' (Kagermann et al., 2013);
- **United Kingdom**: 'Future of Manufacturing' (Foresight, 2013);
- **United States**: 'Advanced Manufacturing Partnership' (President's Council of Advisors on Science and Technology, 2014);
- **France**: 'La Nouvelle France Industrielle' (Conseil National de L'industrie, 2016);
- **European Commission**: 'Factories of the Future' and 'Horizon 2020' (European Commission, 2020);
- **China, Japan and the ROK**: 'Super Smart Society' (Keqiang, 2015);
- **Sweden**: 'Smart Industry' and 'Produktion 2030' (Ministry of Enterprise and Innovation, 2016; Teknikföretagen, 2017)".

1.4 LINKING BETWEEN SM4.0 AND SI4.0

Industry 4.0 is considered one of the most important enablers of achieving manufacturing sustainability (Garbie and Garbie, 2022, 2024). Industry 4.0 technologies can enhance and facilitate implementing manufacturing sustainability in industrial systems. Unfortunately, using Industry 4.0 technologies is still restricted by identifying the requirements for improving sustainability in different levels of manufacturing:

product or machine design, manufacturing process, or manufacturing system level (Enyoghasi and Badurdeen, 2021). The convergence between the digital transformation of manufacturing processes and manufacturing sustainability concepts in industries remains underdeveloped. For these reasons, nowadays several interlinks to join emerging technologies for Industry 4.0 and manufacturing sustainability have been identified (Machado et al., 2020).

The manufacturing sustainability of the current manufacturing systems and enterprises can be improved by adopting these Industry 4.0 technologies (Bai et al., 2020) and enhancing society. It seems that the major aims of adopting and implementing Industry 4.0 technologies in manufacturing systems are not restricted only to traditional performance measurements (time, cost, quality, etc.) but will also contribute to a more comprehensive sustainable performance. The manufacturing sustainability performance is mainly focused on the triple bottom line (TBL), such as economic, social, and environmental. Further understanding of either advanced manufacturing or digital technologies, which are known as emerging technologies for conducting and adopting smart manufacturing and its associated sustainability, is important and urgent for practitioners and academicians.

Understanding and building the relationships between emerging Industry 4.0 technologies and manufacturing sustainability through mathematical modeling is not a trivial task and needs a huge effort from practitioners and academicians. The relationship between the emerging technologies for Industry 4.0 and manufacturing sustainability still needs to be discovered. A more empirical investigation to understand how the emerging technologies for adopting Industry 4.0 are working toward manufacturing sustainability. This means that it is recommended manufacturing companies adopt Industry 4.0 technologies to improve their manufacturing sustainability (Kamble and Gunasekaran, 2019; Bai et al., 2020). The alignment of sustainability in manufacturing enterprise systems (SM4.0) with emerging technologies for Industry 4.0 (I4.0) forms solid evidence of knowledge called a strategy of Sustainability 4.0 (S4.0). This type of strategy is used to support achieving a state of manufacturing sustainability through emerging smart technologies (Reis et al., 2021). In this book, a framework for the alliance of this relationship will be presented and illustrated.

It is important to identify opportunities and barriers for creating sustainability S4.0. The opportunities for researching S4.0 are represented in several issues (Reis et al., 2021). The first issue is the integration between sustainability in manufacturing enterprise systems (SM4.0) and sustainability in the adoption and implementation of Industry 4.0 (S14.0). Therefore, the concept of traditional SCM will be converted to circular SCM or SCM 4.0. The second issue is proposing strategies to integrate SM4.0 with SI4.0 as one of the most important targets. Identifying risk management by implementing S4.0 will be investigated as the third issue. In the fourth issue, the effects of implementing Sustainability 4.0 on the Triple Bottom Line (TBL) will be investigated. Governments, manufacturing industries, and people will also be affected by adopting this implementation. There are some barriers to implementing or creating Sustainability 4.0, represented by high capital costs for the modernization and upgrading of structures and adopting emerging manufacturing and digital technologies; lack of qualified human capital; objectives confusion between SM4.0 and SI4.0; and lack of infrastructure (Reis et al., 2021).

In the manufacturing domain, emerging technologies for Industry 4.0 affect sustainability/sustainable development (S/SD) TBL as much as these TBLs affect Industry 4.0 technologies. Therefore, in answering this comment, "Which comes first, I4.0 or SM4.0?" The answer will be confusing. For this reason, it is recommended to say "Both" but which one can be used to construct and build the second one? It is better to adopt Industry 4.0 to build and construct SM4.0 rather than adopt S/SD to construct Industry 4.0 because Industry 4.0 technologies are considered manufacturing technology, but conducting sustainability practices is a manufacturing philosophy. The impact of Industry 4.0 technologies on manufacturing sustainability has been receiving increasing attention in the last few years. Sustainability and emerging technologies for Industry 4.0 should be considered jointly (Margherita and Braccinil, 2023). Also, a comprehensive evaluation of the relationship between adopting and implementing Industry 4.0 technologies and manufacturing sustainability is missing (Beltrami et al., 2021). Based on Figure 1.1 and the discussion in the previous sections on sustainable manufacturing, Industry 4.0, and the linking between Industry 4.0 and sustainable manufacturing, the Sustainability 4.0 perspectives will be discussed in the following section.

1.5 SUSTAINABILITY 4.0 PERSPECTIVES

Each product or machine design and its associated manufacturing process and industrial system have their identities of structures, like layers and architecture of the whole system. This architecture system is divided into three layers, given, chosen, and core, and as shown in Figure 1.2, it is easy to map them visually. It is important to

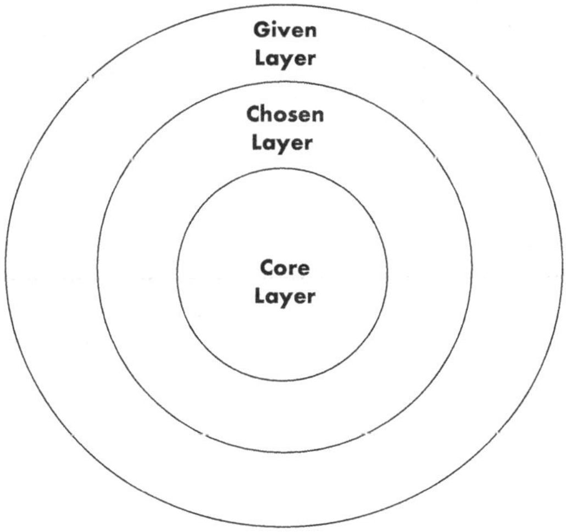

FIGURE 1.2 Layers of architecture system.

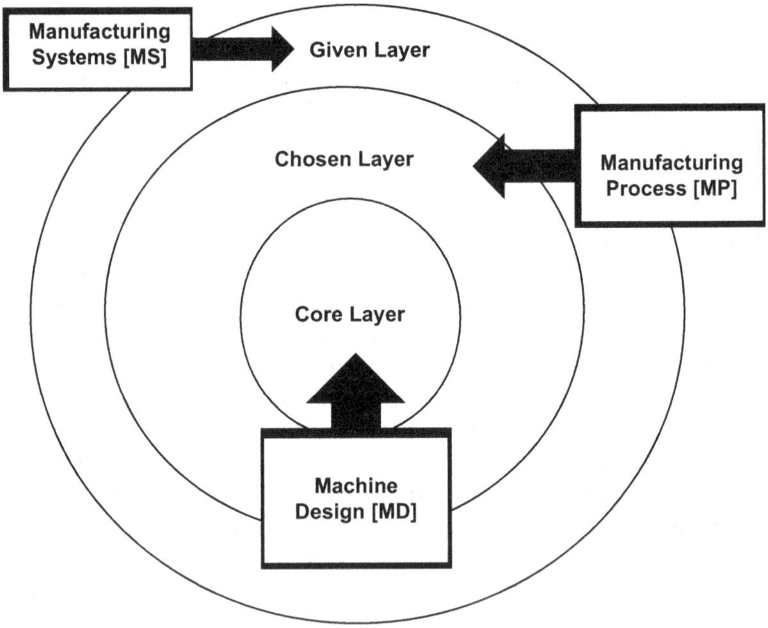

FIGURE 1.3 Architecture layers in sustainable manufacturing.

express the independence and uniqueness of layers without ignoring their foundation and the influence on their components and elements to connect with others.

Most manufacturing products and their manufacturing processes and systems, especially manufacturing industries, are widely focused on these layers. The architecture system is an algorithm connecting layers, and each layer will provide one or more objects to the following layer. Manufacturing, in common, is separated into three primary layers. These layers are represented by machine design (MD) as a core layer, manufacturing processes (MP) as a chosen layer, and manufacturing systems (MS) as a given layer. All these layers constitute manufacturing. Although these layers coincide or overlap at one point, it seems that there is an architectural link between them (Figure 1.3).

Also, an architecture system in Industry 4.0 is just overlapping layers, like sustainable manufacturing, these layers of Industry 4.0/smart manufacturing are connected algorithms that each layer will provide one or more objects to the following layer, like sustainable manufacturing. In this situation, the core is based on a virtual environment, including the advanced emerging technologies or shop floor technologies. This core is used as the basic layer for the chosen layer. The chosen layer mainly identifies which technologies are adopted and selected for the appropriate or feasible manufacturing industry. Lastly, the given layer is the final target of this architecture perspective, which presents and introduces the site readiness for implementation of the plant/factory to conduct Industry 4.0/smart manufacturing (Figure 1.4).

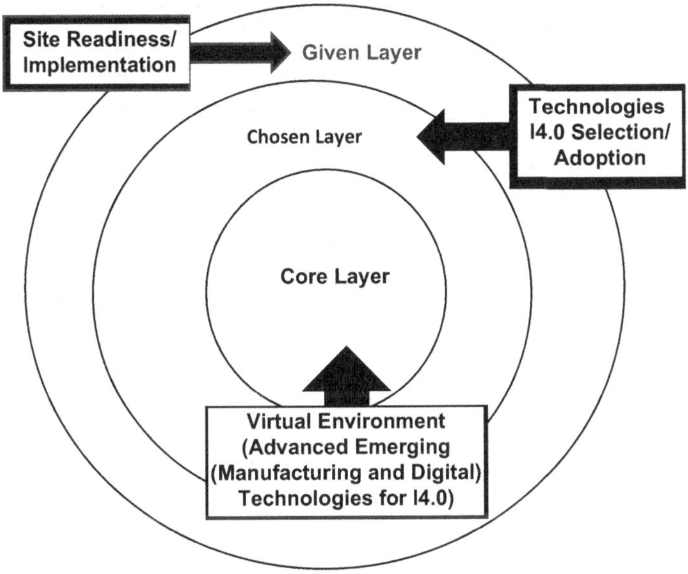

FIGURE 1.4 Architecture layers in Industry 4.0.

1.6 OUTLINE AND SCOPE OF THIS BOOK

The remaining parts of this book are organized as follows. Part 2 discusses and presents the importance of studying and incorporating architecture systems in manufacturing in terms of technical and human perspectives. Part 3 conducts a comprehensive analysis of manufacturing sustainability in either designing the machine and product, or manufacturing processes, or facilities design and layout in terms of types of manufacturing systems from an architecture system. Part 4 portrays an explanatory system for analyzing the architecture systems of Industry 4.0, considering advanced manufacturing technologies and the site readiness of implementing Industry 4.0. Part 5 is used to integrate Industry 4.0 with sustainable manufacturing to distinguish the form of connecting the manufacturing sustainability layers from the Industry 4.0 perspective. Figure 1.5 shows the organization of *Sustainability 4.0: Objectives, Challenges, and Expectations,* which is divided into five main parts and includes 14 chapters.

1.7 CONCLUDING REMARKS

This introductory chapter outlines the key aspects and perspectives explored in the following chapters. It emphasizes the significance of achieving SDGs related to industry and manufacturing. This chapter highlights the role of Industry 4.0 and smart manufacturing in advancing sustainability and explores the critical connection between these concepts. Additionally, it introduces the innovative concept of an architecture perspective as a valuable framework. Finally, this chapter defines the scope of this book, setting the stage for the discussions that follow.

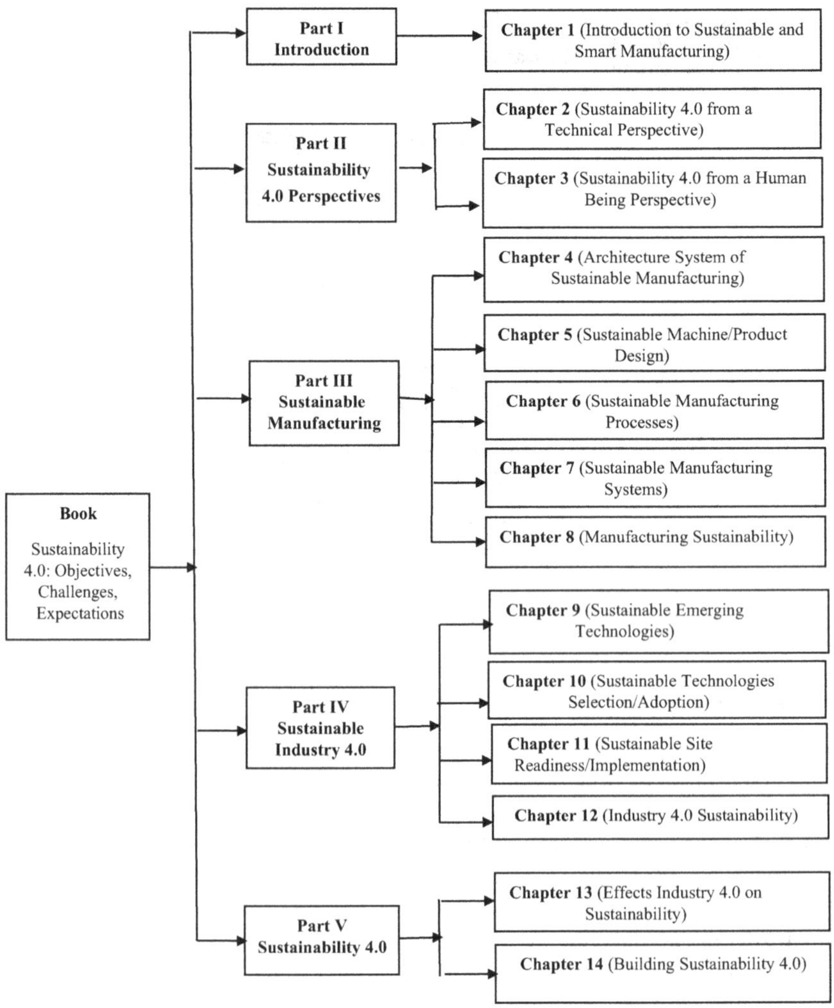

FIGURE 1.5 Organization of this book.

REFERENCES

Bai, C., Dallasega, P., Orzes, G., and Sarkis, J., (2020), "Industry 4.0 Technologies Assessment: A Sustainability Perspective", *International Journal of Production Economics*, Vol. 229, p. 107776.

Beltrami, M., Orzes, G., Sarkis, J., and MarcoSartor, M., (2021), "Industry 4.0 and Sustainability: Towards Conceptualization and Theory", *Journal of Cleaner Production*, Vol. 312, p. 127733.

China, Japan and the ROK, (2015), https://www.mfa.gov.cn/eng/zy/jj/2015zt/lkqdhgjxzsfwb-cxdlczrhldrhy/202406/t20240606_11381625.html

Enyoghasi, C., and Badurdeen, F., (2021), "Industry 4.0 for Sustainable Manufacturing: Opportunities at the Product, Process, and System Levels", *Resources, Conservation, Recycling*, Vol. 166, p. 105362.

European Commission, (2020), Horizon 2020 and the "Factories of the Future" Initiative, https://www.welcomeurope.com/en/horizon-2020-and-the-factories-of-the-future-initiative/.

France, (2016), Rapport annuel 2016 du Conseil national de l'industrie, https://www.entreprises.gouv.fr/la-dge/publications/rapport-annuel-2016-du-conseil-national-de-lindustrie.

Garbie, I., (2013), "DFSME: Design for Sustainable Manufacturing Enterprises (an Economic Viewpoint)," *International Journal of Production Research*, vol. 51, No. 2, pp. 479–503.

Garbie, I., (2014), "An analytical Technique to Model and Assess Sustainable Development Index in Manufacturing Enterprises," *International Journal of Production Research*, Vol. 52, No. 16, pp. 4876–4915.

Garbie, I., and Garbie, A., (2022), "A Holistic Perspective of Sustainable Manufacturing: A Novel Conceptual Approach", *International Journal of Industrial and Systems Engineering*, Vol. 41, No. 2, pp. 135–167.

Garbie, I., and Garbie, A., (2024), "Toward Smart Manufacturing Systems Incorporating Reconfiguration Issues", *International Journal of Industrial and Systems Engineering*, Vol. 46, No. 1, pp. 1–33.

German, (2013), Recommendations for implementing the strategic initiative INDUSTRIE 4.0, https://www.acatech.de/wp-content/uploads/2018/03/Final_report__Industrie_4.0_accessible.pdf

Hasna, A. M., (2010), "Sustainability Classifications in Engineering: Discipline and Approach", *International Journal of Sustainable Engineering*, Vol. 3, No. 4, pp. 258–276.

Kamble, S. S., and Gunasekaran, A., (2019), "Analyzing the Role of Industry 4.0 Technologies and Circular Economy Practices in Improving Sustainable Performance in Indian Manufacturing Organizations", *Production Planning and Control*, Vol. 34, No. 10, pp. 887–901.

Machado, C. G., Winrotha, M. P., and Da Silva, E. H. D. R., (2020), "Sustainable Manufacturing in Industry 4.0: An Emerging Research Agenda", *International Journal of Production Research*, Vol. 58, No. 5, pp. 1462–1484.

Margherita, E. G., and Braccinil, A. M., (2023), "Industry 4.0 Technologies in Flexible Manufacturing for Sustainable Organizational Value: Reflections from a Multiple Case Study of Italian Manufacturers", *Information Systems Frontiers*, Vol. 25, No. 3, pp. 995–1016.

Pansare, R., Yadav, G., Jose Arturo Garza-Reyes, J. A., and Nagare, M. R., (2023), "Assessment of Sustainable Development Goals through Industry 4.0 and Reconfigurable Manufacturing System Practices", *Journal of Manufacturing Technology Management*, Vol. 34 No. 3, pp. 383–413.

Reis, J. S. D. M., Espuny, M., Nunhes, T. V., Sampaio, N. A. D. S., Isaksson, R., De Campos, F. C., and De Oliveira, O. J., (2021), "Striding Towards Sustainability: A Framework to Overcome Challenges and Explore Opportunities through Industry 4.0", *Sustainability*, Vol. 13, pp. 1–28.

Sweden, (2017), Digitizing European Industries - Member States Profile, https://ec.europa.eu/futurium/en/system/files/ged/se_country_analysis.pdf.

United Kindgam, (2013) The Future of Manufacturing: A New Era of Opportunity and Challenge for The UK, https://assets.publishing.service.gov.uk/media/5a7c321140f0b67d0b11f7e6/13-810-future-manufacturing-summary-report.pdf.

United Nations Sustainable Development, (2015), https://sdgs.un.org/goals.

United States, (2014), President's Council of Advisors on Science and Technology (PCAST), https://www.federalregister.gov/documents/2014/12/24/2014-30175/presidents-council-of-advisors-on-science-and-technology-pcast

Yadav, G. Kumar, A., Luthra, S., Garza-Reyes, J. A., Kumare, V., and Batista, L., (2020), "A Framework to Achieve Sustainability in Manufacturing Organizations Developing Economies Using Industry 4.0 Technologies' Enablers", *Computers in Industry*, Vol. 122, 103280.

Part 2

Sustainability 4.0 Perspectives

2 Sustainability 4.0 from a Technical Perspective

2.1 ARCHITECTURAL PERSPECTIVE OF SUSTAINABLE MANUFACTURING

2.1.1 Introduction of Sustainable Manufacturing

The architectural perspective interplay among manufacturing layers suggests that each layer provides elements to the subsequent ones. The core layer—machine design (MD)—supplies components to the intermediate layer—manufacturing processes (MP)—which, in turn, supports the final layer—manufacturing systems (MS). The architectural perspective is a new form of communication to depict the relationship between the hierarchical layers of manufacturing, beginning from the core layer "Machine Design" to the chosen layer "Manufacturing Process" to the given layer "Manufacturing Systems". Although this architectural perspective is like installing a building in nature and unique matter, it can be interpreted within the logical application that is known in this chapter as sustainability in manufacturing enterprise systems. Manufacturing enterprises cover a wide extent of industries such as vehicle and aviation, house things, pharmaceutical, nourishment, computer and electronic gadgets, and drink (Fisher et al., 2018; Garbie and Garbie, 2020a,b,c, 2022, 2024; Garbie et al., 2023). Accomplishing S/SD in manufacturing enterprises could be a sharp concentration on the design of machines (MD), analysis of manufacturing processes (MP), and facilities design and layouts of "manufacturing systems" (MS) (Figure 2.1).

2.1.2 Core Layer: Machine Design (MD)

Machine design (MD) serves as the foundational layer of manufacturing. It involves creating mechanical components such as springs, shafts, gears, bearings, fasteners, and transmission devices like belts and flywheels (Garbie and Garbie, 2022). These components are integral to equipment used in processes like metal forming and machining (e.g., hammers, presses, and machine tools). Ensuring sustainability in machine design involves continuous innovation and updating of components to enhance performance and durability.

2.1.3 Chosen Layer: Manufacturing Processes (MP)

Manufacturing processes (MP) convert raw or semi-finished materials into final products through organized operations. These operations are categorized into solidification, deformation, and machining, with each encompassing various sub-operations such as shaping, property enhancement, and surface preparation. For

DOI: 10.1201/9781003644835-4

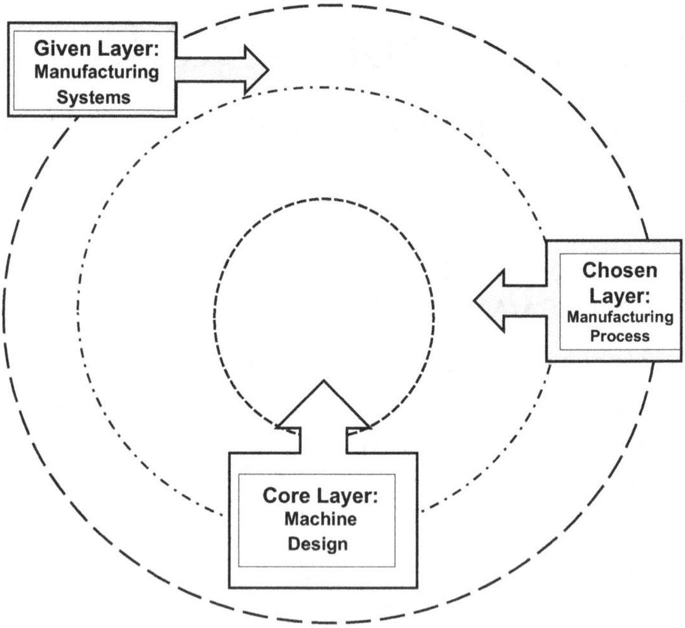

FIGURE 2.1 Layers of sustainable manufacturing regarding architectural perspective.

instance, a mechanical pencil may require ten parts, while an aircraft may involve up to a million. The complexity underscores the importance of systematic approaches in manufacturing processes.

2.1.4 Given Layer: Manufacturing Systems (MS)

As manufacturing systems (MS) comprise numerous assets (people, machines, hardware, etc.) to execute the manufacturing processes (MP) and machine design (MD), they can be called the administration and management of manufacturing processes. For this reason, they are called the given layer since they will be the culmination shape of manufacturing ventures. There are three key system types of manufacturing systems in addition to assembly lines (either manual or robotized). These systems are commonly known as functional or processes layout "Job shop", product layout "flow line", and cellular manufacturing systems (e.g., flexible manufacturing cells and systems). This book aims to analyze and explore the architecture model of the S/SD index for manufacturing.

2.2 ARCHITECTURAL PERSPECTIVE OF INDUSTRY 4.0

2.2.1 Introduction of Industry 4.0

The architectural perspective of Industry 4.0 highlights the interdependence among its layers like sustainable manufacturing in that layers of Industry 4.0 are connected to provide one or more objects. In this situation, the core layer is based on a virtual

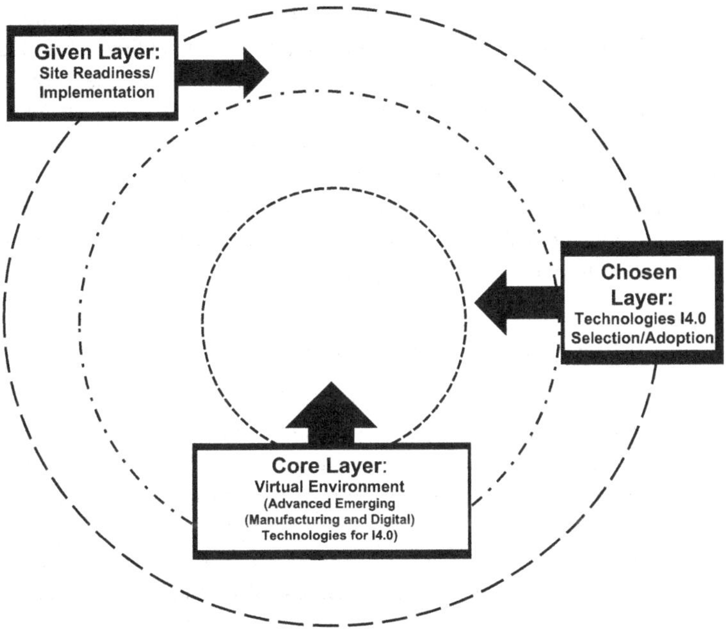

FIGURE 2.2 Layers of Industry 4.0 regarding architectural perspective.

environment, including the emerging manufacturing and digital technologies for Industry 4.0 or the shop floor. The core layer-virtual environment (VE) forms the basis for technology selection (TS), which informs site readiness (SR). This layer is used as the basic layer for the chosen layer. The chosen layer mainly identifies which technologies are selected for the appropriate or feasible manufacturing industry. Lastly, the given layer is the final target of this architectural perspective which presents and introduces the site readiness of the plant/factory to implement Industry 4.0. The architectural perspective is a form of communication to depict the relationship between the hierarchical layers of Industry 4.0 beginning from the core layer "virtual environment for I4.0" to the chosen layer "technologies I4.0 selection" to the given layer "site readiness". It seems that the architectural perspective is like an installing building in nature and a unique matter that can be interpreted within the logical application that is known in this chapter as Industry 4.0 in manufacturing organizations. Accomplishing Industry 4.0/smart manufacturing in industrial organizations could be a sharp concentration on virtual environment (VE) of incorporating emerging digital and manufacturing technologies, technologies I4.0 selection (TS), and site readiness (SR) (Figure 2.2).

2.2.2 Core Layer: Virtual Environment (VE)

The virtual environment (VE) is considered the core and fundamental layer of Industry 4.0. The VE is divided into three sub-layers: emerging or advanced digital and manufacturing technologies; smart storage, data collection, analysis, and sharing technologies; and integration technologies. Each sub-layer has its unique characteristics in terms of the type of technology, and how it can be used and oriented to the manufacturing industry type. Virtual reality (VR), augmented reality (AR), additive manufacturing, the Internet of things, sensors, and robotic systems are classified as emerging advanced manufacturing technologies. Cloud computing (CC), big data analytics (BDA), and the IoT are also identified as smart storage and data collection technologies. Integration technologies are represented in cyber-production systems (CPS) and cyber security systems (CSS). It is important to investigate these enablers of technologies I4.0 when were implemented and adopted. Continuous innovation in these technologies is essential for seamless integration into manufacturing systems.

2.2.3 Chosen Layer: Technologies I4.0 Selection (TS)

Selecting manufacturing technology I4.0 will be based on the nature of the manufacturing industry itself and the different manufacturing processes. Not most of the manufacturing Industry needs the same type of emerging and advanced technologies of I4.0. Each manufacturing has unique requirements and specifications. Technology selection identifies suitable I4.0 technologies for specific manufacturing processes. Not all industries require the same technologies; each has unique specifications. TS bridges the gap between the physical and digital worlds, enabling tailored solutions for diverse manufacturing needs.

2.2.4 Given Layer: Site Readiness (RS)

The given layer is the most important outcome of the previous layers which is used to evaluate and estimate the readiness of the manufacturing organization to adopt the manufacturing technologies for I4.0 and select the appropriate one(s) to be used. For this reason, they are called the given layer since they will be the ultimate estimate of manufacturing ventures in terms of an architectural perspective concerning site readiness. Site readiness evaluates an organization's preparedness to adopt I4.0 technologies. The given layer "SR" represents the final output of the architecture perspective framework, integrating physical and virtual systems to optimize manufacturing processes. SR is a critical measure of a manufacturing enterprise's capability to achieve I4.0.

2.3 CONCLUDING REMARKS

This chapter introduces the architecture perspective approach to sustainable and I4.0/smart manufacturing, emphasizing hierarchical layers. The core, chosen, and given layers for both sustainable manufacturing and sustainable Industry 4.0 are defined, outlining their components and interrelationships. These frameworks set the foundation for detailed exploration in subsequent chapters.

REFERENCES

Fisher, O., Watson, N., Porcu, L., Bacon, D., Rigley, M., and Gomes, R., (2018), "Cloud Manufacturing as a Sustainable Process Manufacturing Route," *Journal of Manufacturing Systems*, Vol. 47, pp. 53–68, 2018.

Garbie, I., and Garbie, A., (2020a), "Sustainability and Manufacturing: A Conceptual Approach", *Proceedings of the Industrial and Systems Engineering Research Conference (IISE Annual Conference and Expo 2020)*, November 1–3, 2020, New Orleans, LA, USA, 6 pp, 2020, Held Virtually.

Garbie, I., and Garbie, A., (2020b), "Outlook of Requirements of Manufacturing Systems for 4.0", *The 3rd International Conference of Advances in Science and Engineering Technology (Multi-Conferences) ASET 2020*, February 4–6, 2020, Dubai, UAE, 2020, 6 pp.

Garbie, I., and Garbie, A., (2020c), "A New Analysis and Investigation of Sustainable Manufacturing through a Perspective Approach", *The 3rd International Conference of Advances in Science and Engineering Technology (Multi-Conferences) ASET 2020*, February 4–6, 2020, Dubai, UAE, 2020, 6 pp.

Garbie, I., and Garbie, A., (2022), "A Holistic Perspective of Sustainable Manufacturing: A Novel Conceptual Approach", *International Journal of Industrial and Systems Engineering*, Vol. 41, No. 2, pp. 135–167, 2022.

Garbie, I., and Garbie, A., (2024), "Toward Smart Manufacturing Systems Incorporating Reconfiguration Issues", *International Journal of Industrial and Systems Engineering*, Vol. 46, No. 1, pp. 1–33, 2024.

Garbie, I., Parsaei, H., and Garbie, M., (2023), "Site Readiness Index for Implementing Industry 4.0 based on Design Principles and Emerging Manufacturing Technologies", *Proceedings of the Industrial and Systems Engineering Research Conference (IISE Annual Conference and Expo 2023)*, May 20–23, 2023, New Orleans, Louisiana, USA, 2023, 6 pp.

3 Sustainability 4.0 from a Human Perspective

3.1 SUSTAINABLE MANUFACTURING

The human-being perspective approach plays a crucial role in sustainable manufacturing by influencing the behavior and decision-making of individuals within the manufacturing process. The human-being perspective approach to sustainable manufacturing has several positive effects on the manufacturing industry. By incorporating human-being perspective principles into adopting sustainable manufacturing practices, manufacturing organizations due to following the human-being perspective system can improve employee motivation, employee engagement, and well-being, leading to increased productivity and job satisfaction.

Additionally, a human-being perspective can help manufacturing organizations better understand consumer behavior and preferences related to sustainable products and design, leading to more effective marketing strategies and increased demand for eco-friendly products. Furthermore, by considering the human perspective impact of sustainable manufacturing on local communities and the environment, manufacturing organizations can develop more ethical and responsible business practices that contribute to long-term sustainability (Garbie and Garbie, 2024).

The human perspective system's effects in sustainable manufacturing affect employee motivation, decision-making, innovation and creativity, behavior changes, and organizational culture. Concerning employee motivation (EM), the human-being perspective approach can impact employee motivation and engagement. Employees are actively more likely to participate in sustainable initiatives. For decision-making (DM), the human-being perspective influences how individuals within the manufacturing layers make decisions related to sustainability. Manufacturing organizations can design interventions to promote sustainable decision-making among employees by understanding cognitive biases, attitudes, and beliefs.

The human-being perspective approach can stimulate innovation and creativity (IC). Encouraging employees to be more creative is one of the most important aspects of the human perspective. Thinking outside the box is one of the significant outcomes of employees bringing new ideas for improving processes, reducing waste, and minimizing environmental impact to enhance sustainability issues in manufacturing. Concerning behavior change (BC), it can facilitate behavior change among employees by promoting awareness of the environmental impact of their actions and encouraging pro-environmental behaviors both at work and at home. For organizational culture (OC), the psychological approach can shape the culture of manufacturing organizations around sustainability within a production facility. Manufacturing organizations can create a culture that supports sustainability issues and practices by promoting values such as environmental responsibility, corporate

 DOI: 10.1201/9781003644835-5

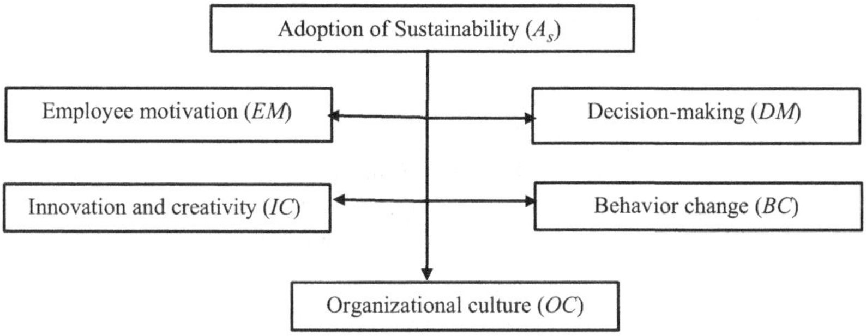

FIGURE 3.1 Issues of adoption sustainability in terms of human-being perspective.

social accountability, and ethical behavior. Overall, the human-being perspective approach has a significant impact on sustainability in manufacturing by influencing individual behavior, decision-making processes, innovation and creativity, organizational culture, and overall commitment within the manufacturing industry.

In general, the human-being perspective approach in sustainable manufacturing can lead to a more holistic and human-centered approach to production that benefits both employees and the environment. This means that the adoption of sustainability concepts (A_S) depends on employee motivation (EM), decision-making (DM), innovation and creativity (IC), behavior change (BC), and organizational culture (OC) (Figure 3.1). The mathematical expression is used to describe this relationship in the following Equation (3.1).

$$A_S(t) = f\big(\mathrm{EM}(t),\, \mathrm{DM}(t),\, \mathrm{IC}(t),\, \mathrm{BC}(t),\, \mathrm{OC}(t)\big) = \left\{ \begin{bmatrix} \mathrm{EM}(t) \\ \mathrm{DM}(t) \\ \mathrm{IC}(t) \\ \mathrm{BC}(t) \\ \mathrm{OC}(t) \end{bmatrix} \right\} \tag{3.1}$$

Equation (3.1) is rewritten as Equations (3.2) and (3.3)

$$A_S(t) = \sum_{i=1}^{j=5} W_{ij}(t)\; X_i(t) \tag{3.2}$$

$$\begin{aligned} A_S(t) = {} & W_{\mathrm{EM}}(t)\; \mathrm{EM}(t) + W_{\mathrm{DM}}(t)\mathrm{DM}(t) + W_{\mathrm{IC}}(t)\mathrm{IC}(t) \\ & + W_{\mathrm{BC}}(t)\mathrm{BC}(t) + W_{\mathrm{OC}}(t)\mathrm{OC}(t) \end{aligned} \tag{3.3}$$

where:

$A_S(t)$ = adoption level or index of sustainability regarding human-being perspective at time t,

$X_i(t)$ = adopting effect (i) at time t.

$$X_i(t) = \frac{E^i - I^i_{\min}}{I^i_{\max} - I^i_{\min}} \tag{3.4}$$

E^i = percentage of existing value of the effect (i) at time t,

$I^i_{\max}$ = maximum percentage of the effect (i) at time t,

$I^i_{\min}$ = minimum percentage of the effect (i) at time t.

$\mathrm{EM}(t)$ = Employees' motivation level or index regarding sustainability at time t.

$$\mathrm{EM}(t) = \frac{E^{\mathrm{EM}} - I^{\mathrm{EM}}_{\min}}{I^{\mathrm{EM}}_{\max} - I^{\mathrm{EM}}_{\min}} \tag{3.5}$$

E^{EM} = percentage of the existing employees' motivation regarding sustainability at the time t,

$I^{\mathrm{EM}}_{\max}$ = maximum percentage of the employees' motivation regarding sustainability at the time t,

$I^{\mathrm{EM}}_{\min}$ = minimum percentage of the employees' motivation regarding sustainability at the time t,

$\mathrm{DM}(t)$ = decision-making level or index regarding sustainability at the time t.

$$\mathrm{DM}(t) = \frac{E^{\mathrm{DM}} - I^{\mathrm{DM}}_{\min}}{I^{\mathrm{DM}}_{\max} - I^{\mathrm{DM}}_{\min}} \tag{3.6}$$

E^{DM} = percentage of the existing decision-making regarding sustainability at the time t,

$I^{\mathrm{DM}}_{\max}$ = maximum percentage of the decision-making regarding sustainability at the time t,

$I^{\mathrm{DM}}_{\min}$ = minimum percentage of the decision-making regarding sustainability at the time t,

$\mathrm{IC}(t)$ = innovation and creativity level or index regarding sustainability at the time t.

$$\mathrm{IC}(t) = \frac{E^{\mathrm{IC}} - I^{\mathrm{IC}}_{\min}}{I^{\mathrm{IC}}_{\max} - I^{\mathrm{IC}}_{\min}} \tag{3.7}$$

E^{IC} = percentage of the existing innovation and creativity regarding sustainability at the time t,

$I^{\mathrm{IC}}_{\max}$ = maximum percentage of the innovation and creativity regarding sustainability at the time t,

$I^{\mathrm{IC}}_{\min}$ = minimum percentage of the innovation and creativity regarding sustainability at the time t,

$\mathrm{BC}(t)$ = behavior change level or index of employees regarding sustainability at the time t.

$$\mathrm{BC}(t) = \frac{E^{\mathrm{BC}} - I^{\mathrm{BC}}_{\min}}{I^{\mathrm{BC}}_{\max} - I^{\mathrm{BC}}_{\min}} \tag{3.8}$$

E^{BC} = percentage of the existing behavior change regarding sustainability at the time t,

I_{max}^{BC} = maximum percentage of the behavior change regarding sustainability at the time t,

I_{min}^{BC} = minimum percentage of the behavior change regarding sustainability at the time t,

$OC(t)$ = organizational culture level or index regarding sustainability at the time t.

$$OC(t) = \frac{E^{OC} - I_{min}^{OC}}{I_{max}^{OC} - I_{min}^{OC}} \tag{3.9}$$

E^{OC} = percentage of the existing organizational culture regarding sustainability at the time t,

I_{max}^{OC} = maximum percentage of the organizational culture regarding sustainability at the time t,

I_{min}^{OC} = minimum percentage of the organizational culture regarding sustainability at the time t.

The symbols $W_{EM}(t)$, $W_{DM}(t)$, $W_{IC}(t)$, $W_{BC}(t)$, and $W_{OC}(t)$ are the relative weights for the employee's motivation, decision-making, innovation and creativity, behavior change, and organizational culture regarding sustainable development at time t for psychological approach, respectively. The analytical hierarchy process (AHP) will be used to identify the relative weights between those effects in the following matrix (Garbie, 2013, 2014).

3.2 INDUSTRY 4.0

Adopting Industry 4.0 or smart manufacturing involves a significant shift in how businesses operate and the technologies they use. This shift can profoundly impact the human-being perspective approach of individuals within an organization, including employees, managers, and leaders.

One key aspect of the human-being perspective system in adopting Industry 4.0 is resistance to change. Resistance to change is one form of people following their beliefs and methodologies of work. It can generate a big challenge for them in terms of existence. In the context of Industry 4.0, this resistance can manifest as fear of job loss due to automation, skepticism about the benefits of new technologies, or reluctance to learn new skills. To address this resistance, organizations need to focus on communication and education. Leaders should communicate the reasons for adopting Industry 4.0, the benefits it will bring to the organization and its employees, and how individuals can adapt and thrive in this new environment. Providing training and support for employees to develop new skills and competencies is also crucial in overcoming resistance to change (Garbie and Garbie, 2024).

Another important aspect of the human-being perspective system in adopting Industry 4.0 is the mindset shift. Moving toward smart manufacturing requires a shift from a traditional mindset focused on efficiency and cost reduction to a more innovative mindset that embraces technology, collaboration, and continuous improvement.

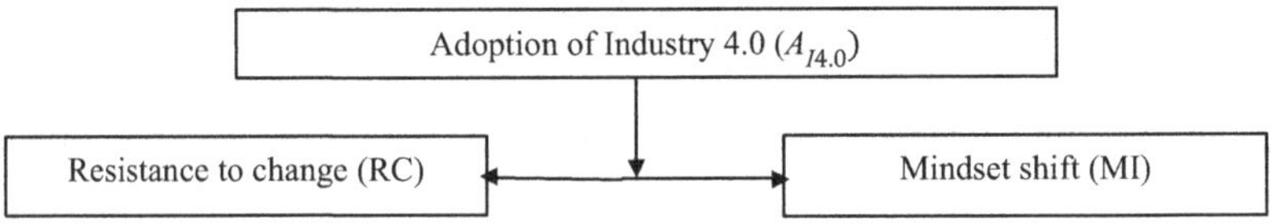

FIGURE 3.2 Issues of adoption Industry 4.0/in terms of human-beings perspective.

This shift can be challenging for individuals who are used to working in a certain way or who are resistant to new ideas. Organizations can support this mindset shift in several forms such as innovation and creation, experimentation, problem-solving, and risk management. Leaders should be open and supportive of new ideas with initiative and innovation and give another opportunity if there are failures.

Overall, the human-being perspective system in adopting Industry 4.0 is complex and multifaceted. By addressing resistance to change, fostering a mindset shift toward innovation, and providing support for employees to develop new skills, organizations can successfully navigate this transition and unlock the full potential of the adoption Industry 4.0.

This means that the adoption of Industry 4.0 ($A_{I4.0}$) in terms of human being's perspective depends on resistance to change (RC) and mindset shift (MI) (Figure 3.2). The mathematical expression is used to describe this relationship in the following Equation (3.10).

$$A_{I4.0}(t) = f\left(\text{RC}(t),\ \text{MI}(t)\right) = \left\{ \begin{array}{c} \text{RC}(t) \\ \text{MI}(t) \end{array} \right\} \tag{3.10}$$

Equation (3.10) can be written as Equations (3.11) and (3.12)

$$A_{I4.0}(t) = \sum_{i=1}^{j=2} W_{ij}(t)\ X_i(t) \tag{3.11}$$

$$A_{I4.0}(t) = W_{\text{RC}}(t)\ \text{RC}(t) + W_{\text{MI}}(t)\text{MI}(t) \tag{3.12}$$

where:

$A_{I4.0}(t)$ = adoption level or index of Industry 4.0 based on human-being perspective at time t,

$X_i(t)$ = the adopting aspect (i) at any time t.

$$X_i(t) = \frac{E^i - I^i_{\min}}{I^i_{\max} - I^i_{\min}} \tag{3.13}$$

E^i = percentage of using existing value of the adopting aspect (i) at time t,

$I^i_{\max}$ = maximum percentage of using the adopting aspect (i) at time t,

$I^i_{\min}$ = minimum percentage of using the adopting aspect (i) at the time,

$\text{RC}(t)$ = resistance level or index of people regarding adopting Industry 4.0 at time t.

$$\mathrm{RC}(t)=\frac{E^{\mathrm{RC}}-I_{\min}^{\mathrm{RC}}}{I_{\max}^{\mathrm{RC}}-I_{\min}^{\mathrm{RC}}} \tag{3.14}$$

E^{RC} = percentage of using the existing value of people's resistance to change of adopting Industry 4.0 at the time t,

$I_{\max}^{\mathrm{RC}}$ = maximum percentage of people's resistance to change of adopting Industry 4.0 at the time t,

$I_{\min}^{\mathrm{RC}}$ = minimum percentage of using of people's resistance to change of adopting Industry 4.0 at any time t,

$\mathrm{MI}(t)$ = level or index of mindset shift of employees in manufacturing enterprises system regarding adopting Industry 4.0 at time t.

$$\mathrm{MI}(t)=\frac{E^{\mathrm{MI}}-I_{\min}^{\mathrm{MI}}}{I_{\max}^{\mathrm{MI}}-I_{\min}^{\mathrm{MI}}} \tag{3.15}$$

E^{MI} = percentage of existing employees' mindset shifted of adopting Industry 4.0 at any time t,

$I_{\max}^{\mathrm{MI}}$ = maximum percentage of employees' mindset shifted of adopting Industry 4.0 at any time t,

$I_{\min}^{\mathrm{MI}}$ = minimum percentage of employees' mindset shifted of adopting Industry 4.0 *at any time t.*

The symbols $W_{\mathrm{RC}}(t)$, $W_{\mathrm{MI}}(t)$ are the relative weights for the people's resistance to change and mindset shift of adopting Industry 4.0 at time t, respectively.

3.3 NUMERICAL EXAMPLE 3.1

This numerical example illustrates the proposed methodology for evaluating the adoption of the sustainability index in manufacturing industries concerning the human perspective in terms of human beings. The key indicators related to sustainability issues and performance metrics, including existing values and target ranges (minimum and maximum), are summarized in Table 3.1.

Concerning the issues of the adoption of sustainability in terms of human beings (Table 3.1), the foremost imperative issue is normalizing these issues of sustainability through their indicators. For example, the indicator for employee motivation (EM) is evaluated as the percentage of motivation at a given time t by utilizing two values for the target (greatest as 100% and least as 80%) and one existing value as 85%. At that point, the normalized value of motivation is assessed as 0.25 as given in the following equation and appears in Table 3.2. The total assessed indicators result in issues of the adoption of sustainability in terms of human beings, which appears in Table 3.2.

$$\mathrm{EM}(t)=\frac{85-80}{100-80}=0.25$$

The following AHP matrix, as shown in Figure 3.3, illustrates the relative weights between the issues of adoption sustainability. It can be noticed from the AHP matrix

TABLE 3.1
Issues of Adoption of Sustainable Manufacturing in Terms of Human Perspective

Indicator	Units Performance Measures (%)	Performance Metrics Existing Value	Target Value Max	Min
EM	Percentage (%) of motivating employees at any time t	85	100	80
IC	Percentage (%) of using innovation and creativity at any time t	85	90	80
DM	Percentage (%) of conducting decision-making at any time t	95	100	90
BC	Percentage (%) of changing behavior at any time t	20	30	15
OC	Percentage (%) of accepting organizational culture (OC) at any time t	85	100	80

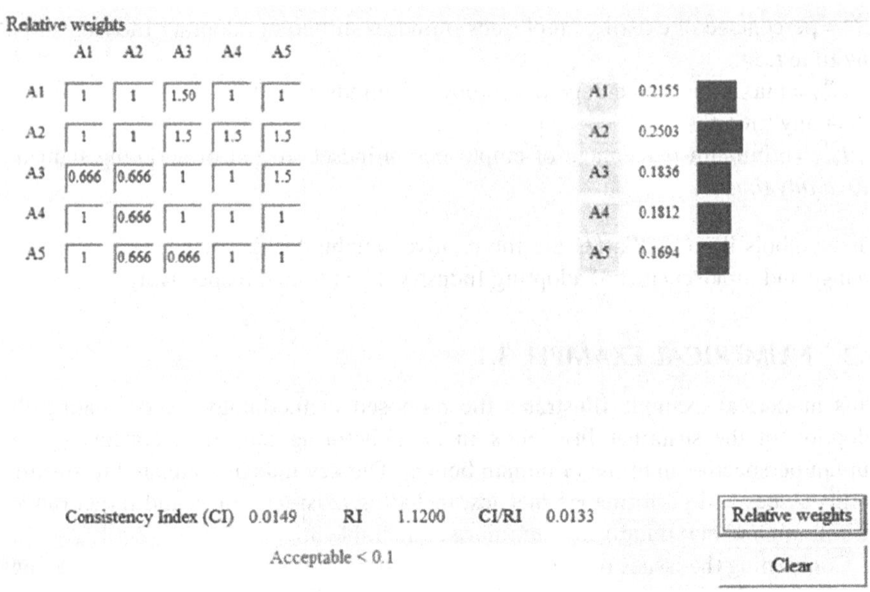

FIGURE 3.3 Estimation of the relative weights between issues of adopting sustainability.

TABLE 3.2
Adoption Index in Sustainable Manufacturing in Terms of Human Beings

Indicator	EM	IC	DM	BC	OC
Normalized index	0.25	0.50	0.50	0.33	0.25
Relative weights	0.2155	0.2503	0.1836	0.1812	0.1694
$A_S(t)$	0.3730				

based on stakeholders' estimation that the employees' motivation (EM) was estimated to be equivalent important to innovation and creation (IC), behavior change (BC), and organizational culture (So5) and one and a half more important than decision-making (DM). The IC was estimated to be one and a half more important than DM, BC, and OC. Also, the DM was estimated to be equivalent to BC and is estimated to be one and a half more important than OC. The So*4* is estimated to be equivalent important to OC. As a result, the relative weights are estimated for the issues of adoption sustainability model as (0.2155, 0.2503, 0.1836, 0.1812, and 0.1694) for employee motivation, innovation, and creation, decision-making, behavior changes, and organizational culture, respectively (Figure 3.3). Thus, the adoption sustainability model Equation (3.3) is calculated, and the final result appears in Table 3.2.

$$A_S = \begin{array}{c} \text{EM} \\ \text{IC} \\ \text{DM} \\ \text{BC} \\ \text{OC} \end{array} \begin{bmatrix} 1 & 1 & 1.5 & 1 & 1 \\ 1 & 1 & 1.5 & 1.5 & 1.5 \\ 1/1.5 & 1/1.5 & 1 & 1 & 1.5 \\ 1 & 1/1.5 & 1 & 1 & 1 \\ 1 & 1/1.5 & 1/1.5 & 1 & 1 \end{bmatrix}$$

The index of adopting sustainability at a given time *t*, $cA_S(t)$, is evaluated by utilizing Equation (3.3). The results are outlined in Table 3.2 as 0.3730 (37.30%). The index of adopting sustainability breaks even with 37.30%, and this value appears generally little, it must be moved forward in all indicators of the adoption of sustainable manufacturing in terms of human beings.

For the adoption and implementation of Industry 4.0, the distinguishing indicators of adopting Industry 4.0 issues and performance measurements (existing values and target values (minimum and maximum)) are outlined in Table 3.3.

The normalized values of Industry 4.0 issues are assessed in Table 3.4. In Table 3.4, the total assessed indicators result in issues of the adoption of Industry 4.0 in terms of human beings, and the relative weights between the resistance of employees and

TABLE 3.3
Issues of Adoption Industry 4.0/Smart Manufacturing in Terms of Human Beings

		Performance Metrics		
	Units	Existing	Target Value	
Indicator	Performance Measures (%)	Value	Max	Min
RC	Percentage (%) of employees resistant to change at any time *t*	30	40	20
MI	Percentage (%) of employees' mindset shifted at any time *t*	50	70	40

TABLE 3.4
Adoption Index for Industry 4.0 in Terms of Human Beings

Indicator	EM	IC
Normalized Index	0.50	0.33
Relative Weights	0.50	0.50
$A_{I,4.0}(t)$	0.4150	

mindset shift are estimated as equivalent importance. The index of adopting Industry 4.0/smart manufacturing at any time t, $A_{I4.0}(t)$, is evaluated by utilizing Equation (3.12). The results are outlined in Table 3.4 as 0.4150 (41.50%). The index of adopting Industry 4.0/smart manufacturing breaks even with 41.50%, and this value appears close to medium.

3.4 CONCLUDING REMARKS

This chapter proposed and analyzed the human-being perspective approach to adopting either Sustainable Manufacturing (SM4.0) or Industry 4.0/Smart Manufacturing (I4.0) using a holistic and human-centered perspective. Several personality-related factors were examined for each approach. The concepts underlying these factors, along with the components specific to Sustainable Manufacturing and Industry 4.0/ Smart Manufacturing, were mathematically formulated and expressed to translate qualitative aspects into quantitative metrics.

REFERENCES

Garbie, I., (2013), "DFSME: Design for Sustainable Manufacturing Enterprises (An Economic Viewpoint)," *International Journal of Production Research*, Vol. 51, No. 2, pp. 479–503.

Garbie, I., (2014), "An Analytical Technique to Model and Assess Sustainable Development Index in Manufacturing Enterprises," *International Journal of Production Research*, Vol. 52, No. 16, pp. 4876–4915.

Garbie, I., and Garbie, M., (2024), "Sustainability of Manufacturing Enterprises: As Psychological Perspective of Linking Manufacturing Layers", *Proceedings of the Industrial and Systems Engineering Research Conference (IISE Annual Conference and Expo 2024)*, May 18–21, 2024, Montreal, Canada, 6 pp.

Part 3

Sustainable Manufacturing

4 Architecture System of Sustainable Manufacturing

4.1 INTRODUCTION AND BACKGROUND

The sustainability of manufacturing industries should consider all issues of expansion in terms of resources (e.g., people, machines, material handling equipment, etc.). Achieving sustainable development in manufacturing industries requires full consideration of designing machines and products, analysis and investigation of manufacturing processes, and finally the facilities design and layout systems (Garbie, 2013, 2014). Manufacturing sustainability is considered to be an untraditional concept to be used as a new performance measurement regarding the survival of manufacturing industries to deal with any domestic and international financial crisis.

As manufacturing sustainability is mainly composed of three different layers as mentioned in the above paragraph, studying sustainable manufacturing is urgent, especially investigating each manufacturing layer first individually (Garbie and Garbie, 2022). Manufacturing is mainly defined as relying on economic concepts and technological advancement; the architecture system perspective is still unexplored in most of the research works (Garbie and Garbie, 2022, 2024). The problem addressed in this part (Part 2), in general, and this chapter specifically, is how to depict manufacturing sustainability to guide manufacturing industries toward what they need to be sustainable and what the requirements are to be achieved. Figure 4.1 illustrates the concepts of manufacturing sustainability with major layers of manufacturing.

Until presently as it was noticed, a few distributed published works have been utilized within the area of manufacturing sustainability, but none of the work specified the manufacturing sustainability in each layer either in designing machine or product or manufacturing process or facilities design and layout "manufacturing systems" (Garbie, 2014; Garbie and Garbie, 2022, 2024; Garbie et al., 2023). A modular machine component as an enabler for reconfigurable machine apparatuses through machining forms in CNC vertical milling machines was recommended (Shneor, 2018). Moreover, the basic characteristics, impediments, and opportunities for utilizing cloud manufacturing toward sustainability in a discrete manufacturing process were examined (Fisher et al., 2018). They specified common concepts/issues of sustainability without accentuation on machining processes. An axes adaptive compensation table was designed, manufactured, and tested to update the usefulness of processing machines (Kianinejad et al., 2018). Already published papers on the sustainability in machining processes through four fundamental issues: design, modeling, simulation, and optimization and appraisal were surveyed (Alvarez et al., 2017).

DOI: 10.1201/9781003644835-7

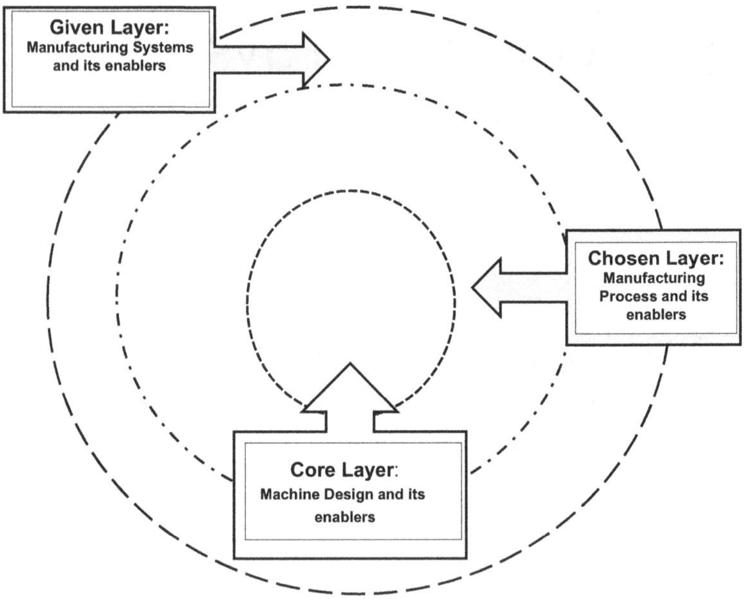

FIGURE 4.1 Manufacturing sustainability with its layers.

Cryogenic machining as a sustainable solution and/or elective instead of flood cooling in terms of flow quality, thermal insulation, controllability, and safety to minimize or dispose of the unfavorable impacts was proposed (Goindi and Sarkar, 2017; Lu et al., 2018). They connected three proposed approaches through a case study on a titanium amalgam through distinctive sorts of processing operations. The impact of coolants/lubrications on machining processes such as dry, flood, and water vapor and their impact on material removal rate (MRR) to realize sustainability taken into thought for the machining environment (administrator well-being and security, outflows, reusing, and transfer) was displayed (Kadam and Pawade, 2017). The machined part (portion cleaning, scrap, transfer, surface unpleasantness, and remaining stresses) was also illustrated (Kadam and Pawade, 2017).

A comprehensive writing survey concerning sustainability in manufacturing ventures concerning the manufacturing system was conducted (Garbie, 2013). A global reporting initiative (GRI) was used through standards and techniques to the examination of the reports of sustainability distributed by multinational organizations (Garbie and Garbie, 2022). A system to assess the manufacturing sustainability of operations was proposed (Garbie and Garbie, 2024). Reconfiguration is one of the most important forms to achieve manufacturing sustainability (Garbie and Garbie, 2024).

4.2 MANUFACTURING SUSTAINABILITY LAYERS

The primary objective of this chapter is to analyze manufacturing sustainability by identifying sustainability enablers within each manufacturing layer. Numerical models are proposed to evaluate sustainability indexes of designing machines and products, analysis and investigation of manufacturing processes, and facilities design and layout "manufacturing systems". Figure 4.2 illustrates the elements and components of each manufacturing layer.

4.2.1 Machine Component Design (Core Layer)

This foundational layer is divided into two main areas: initial machine design and retrofitting/upgrading.

1. **Initial Machine Design**:
    - Key elements include shafts, gears, bearings, spindles, screws/bolts, power transmission systems, and snap fittings.
    - Mechanical properties such as tensile strength, compression, impact, shearing, and torsion are critical considerations.
2. **Retrofitting/Upgrading**:
    - Recommended components include modular modules, interchangeable tool heads, frames, tables, and Internet of Things (IoT) integration to enhance machine capabilities.

4.2.2 Manufacturing Processes (Chosen Layer)

This layer focuses on three primary manufacturing processes:

1. **Solidification Processes**:
    - Includes operations like sand casting and die casting.
2. **Deformation Processes**:
    - Covers rolling, forging, extrusion, drawing, and sheet metal working operations.
3. **Machining Processes**:
    - Encompasses conventional operations (e.g., turning, milling, drilling, shaping, grinding) and non-conventional machining methods.

4.2.3 Manufacturing Systems (Given Layer)

This layer represents the overarching structure of manufacturing and includes four main system types:

1. **Job Shop Manufacturing Systems**:
    - Suitable for low-volume, high-variety production.
2. **Cellular (Focused) Manufacturing Systems**:
    - Optimized for efficiency in batch production.
3. **Flow Line Manufacturing Systems**:
    - Used for high-volume, low-variety production.
4. **Assembly Lines**:
    - Can be manual or automated, depending on production requirements.

By addressing these layers—design of machine and product components, analysis of manufacturing processes, and facilities design and layout, this chapter provides a comprehensive framework for understanding and implementing manufacturing sustainability.

4.3 THE TRIPLE BOTTOM LINE OF MANUFACTURING SUSTAINABILITY

Sustainability (S4.0) is poised to drive the 21st century in much the same way that automation shaped the 20th century and steam power revolutionized the 19th century. Among these pillars of manufacturing sustainability, this chapter focuses on integrating sustainable manufacturing (SM 4.0) with the architecture perspective into the TBL framework, highlighting the benefits of each pillar and their interactions with one another (Figure 4.3).

It can be noticed from Figure 4.3 that it illustrates the long-term features associated with each pillar. The interaction between these pillars generates additional benefits, including short-term features that arise from their interactions. There are a lot of short-term features representing job creation, enhancement in skills, local economic impacts, social investments, business ethics, and taxes/loyalties. At the same time, short-term benefits/features are generated due to integrating social sustainability and environmental sustainability like safety and health, global climate change, local environment impact, and resource management. Life cycle analysis, product stewardship, and resource efficiency are the short-term features of integrating economic sustainability and environmental sustainability. These features and their associated benefits are measured by using sustainable indicators with their performance

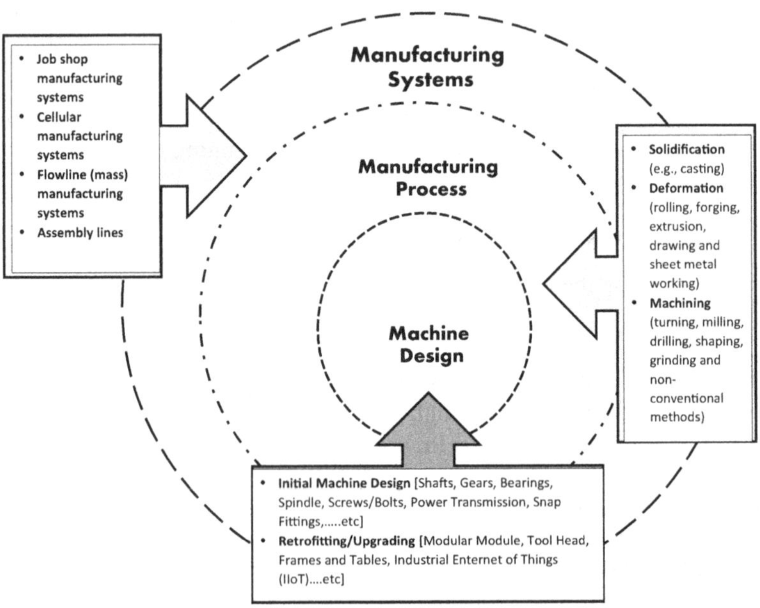

FIGURE 4.2 Layers of sustainable manufacturing in terms of architecture perspectives.

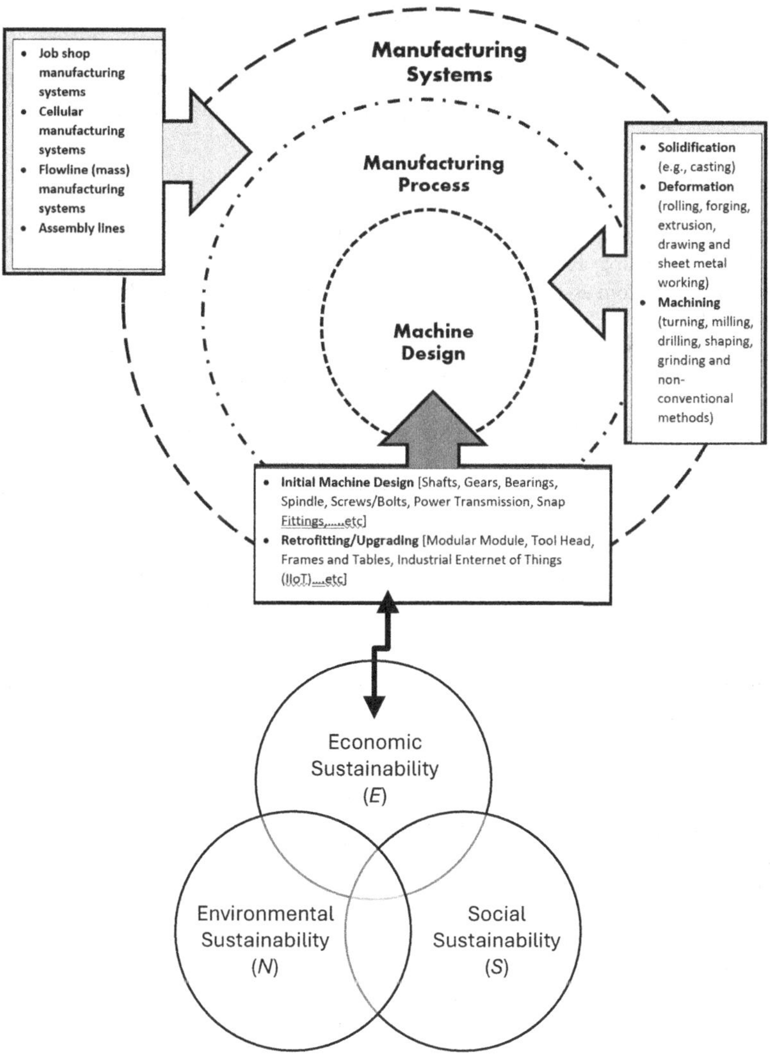

FIGURE 4.3 Integrating the triple bottom line (TBL) of sustainability with the architecture perspective of sustainable manufacturing.

measurements which are so-called key performance indicators (KPI). These indicators will be discussed and explained with each layer of sustainable manufacturing in the following Chapters 5–7.

4.4 CONCLUDING REMARKS

In this chapter, the three layers of manufacturing sustainability, examined from the architectural perspective, were proposed and analyzed in conjunction with the Triple Bottom Line (TBL). The concepts of these layers were introduced along with their respective sustainability enablers. This chapter also discussed the features of TBL, highlighting both long-term and short-term benefits. Additionally, a framework for assessing manufacturing sustainability indexes was presented, providing a structured approach to measure and evaluate progress in sustainable manufacturing.

REFERENCES

Alvarez, M. E. P., Barcena, M. M., and Gonzalez, F. A., (2017), "On the sustainability of Machining Process, Proposal for a Unified Framework through the Triple Bottom-Line from an Understanding Review", *Journal of Cleaner Production*, Vol. 142, pp. 3890–3904.

Fisher, O., Watson, N., Porcu, L., Bacon, D., Rigley, M., and Gomes, L. G., (2018), "Cloud Manufacturing as a Sustainable Process Manufacturing Route", *Journal of Manufacturing Systems*, Vol. 47, pp. 53–68.

Garbie, I., (2013), "DFSME: Design for Sustainable Manufacturing Enterprises (An Economic Viewpoint)", *International Journal of Production Research*, Vol. 51, No. 2, pp. 479–503.

Garbie, I., (2014), "An Analytical Technique to Model and Assess Sustainable Development Index in Manufacturing Enterprises", *International Journal of Production Research*, Vol. 52, No. 16, pp. 4876–4915.

Garbie, I., Garbie, A., (2022), "A Holistic Perspective of Sustainable Manufacturing: A Novel Conceptual Approach", *International Journal of Industrial and Systems Engineering*, Vol. 41, No. 2, pp. 135–167.

Garbie, I., Garbie, A., (2024a), "Toward Smart Manufacturing Systems incorporating Reconfiguration Issues", *International Journal of Industrial and Systems Engineering*, Vol. 46, No. 1, pp. 1–33.

Garbie, I., and Garbie, M., (2024b), "Sustainability of Manufacturing Enterprises: As Psychological Perspective of Linking Manufacturing Layers", *Proceedings of the Industrial and Systems Engineering Research Conference (IISE Annual Conference and Expo 2024)*, May 18–21, 2024, Montreal, Canada, 6 pp.

Garbie, I., Parsaei, H., and Garbie, M., (2023), "Site Readiness Index for Implementing Industry 4.0 based on Design Principles and Emerging Manufacturing Technologies", *Proceedings of the Industrial and Systems Engineering Research Conference (IISE Annual Conference and Expo 2023)*, May 20–23, 2023, New Orleans, Louisiana, USA, 2023, 6 pp.

Goindi, G. S., and Sarkar, P., (2017), "Dry Machining: A Step towards Sustainable Machining-Challenges and Future Directions", *Journal of Cleaner Production*, Vol. 165, pp. 1557–1571.

Kadam, G. S., and Pawade, R. S. (2017), "Surface Integrity and Sustainability Assessment in High-Speed Machining of Inconel 718- An Eco-Friendly Green Approach", *Journal of Cleaner Production*, vol. 147, pp. 273–283.

Kianinejad, K., Thorn, S., Kushwaha, S., and Uhlmann, E., (2018), "Add-on Error Compensation Unit as Sustainable Solution for Outdated Milling Machines", *Procedia CIRP*, Vol. 40, pp. 174–178.

Lu, T., Kudaravalli, R., and Georgiou, G., (2018), "Cryogenic Machining through the Spindle and Tool for Improved Machining Process Performance and Sustainability: Pt. I, Systems Design", *Procedia Manufacturing*, Vol. 21, pp. 266–272.

Shneor, Y., (2018), "Reconfigurable Machine Tool: CNC Machine for Milling Grinding and Polishing", *Procedia Manufacturing*, Vol. 21, pp. 221–227.

5 Sustainable Machine/ Product Design

5.1 INTRODUCTION TO MACHINE DESIGN (MD)

The design of manufacturing enterprise systems is increasingly influenced by rapidly changing market conditions and the need to develop both new and existing products. Central to these systems are machines and equipment, particularly machine tools. These machine tools, "cutting tools", require specific characteristics in both hardware and software design to meet modern new products and fluctuating forecasting demands.

Historically, machine design has evolved with the Industrial Revolution. Machines powered by steam defined Industry 1.0, while Industry 2.0 introduced electromechanical equipment. Industry 3.0 saw the integration of automation and computer connectivity. In the ongoing Industry 4.0 era, machine design is shifting toward integration with Cyber-Physical Production Systems (CPPSs) and Industrial Internet of Things (IIoT) networks (Garbie and Parsaei, 2022). However, the full realization of machine design for Industry 4.0 is still emerging.

A new generation of machine design is urgently required to align with Industry 4.0 objectives. These designs must incorporate advanced technologies to support sustainable manufacturing practices and the transition to smart manufacturing systems. To meet these needs, the interdisciplinary field of mechatronics—integrating mechanics, electronics, and computer science—plays a pivotal role. It facilitates the development of sustainable machine designs by leveraging information technology (IT) and digital technology (DT). These technologies enable data capture during manufacturing processes, establishing smart networks that link the real and virtual worlds through IT, DT, and IIoT.

While conventional machines continue to be widely used in many factories globally, their suitability for Industry 4.0 is limited. Upgrading these machines to adopt advanced manufacturing technologies can enhance their efficiency and enable their integration into smart manufacturing systems. However, Computerized Numerical Control (CNC) machines remain the preferred option for Industry 4.0 due to their modularity, flexibility, and ability to handle complex and specific operations. CNC machines are designed to accommodate changes in predictive demand while seamlessly integrating with advanced manufacturing technologies and information systems.

Sustainability in machine design is a foundational element of sustainable manufacturing, extending to manufacturing processes and systems (Garbie and Garbie, 2022, 2024a,b). Modular machine components, such as reconfigurable apparatuses for CNC vertical milling machines, have been recommended as sustainability enablers (Shneor, 2018). For example, an adaptive compensation table was designed and tested to enhance the functionality of processing machines (Kianinejad et al., 2018).

DOI: 10.1201/9781003644835-8

5.2 MACHINE/PRODUCT DESIGN WITHIN AN ARCHITECTURE PERSPECTIVE

The concepts of machine design are based on the designing of components and parts of the machines either fixed or movable (horizontal or vertical) or rotational. Machine design focuses on the creation and optimization of mechanical components, including springs, shafts, gears, bearings, fasteners, joining mechanisms, and power transmission devices (e.g., belts and flywheels). These types of parts and components will be classified in Figure 5.1. These components form the foundation of the machine's functionality and play a critical role in achieving manufacturing sustainability objectives

5.2.1 Sustainable Machine Design Challenges and Solutions

Sustainable machines, particularly machine tools, represent a critical challenge in modern manufacturing. The manufacturers of machine components and their corresponding machines play a significant role in designing sustainable production machine tools, focusing on key factors such as reliability and availability, efficiency and effectiveness, environment and userfriendliness, and ergonomics (Garbie and Garbie, 2024). The machine tools industry, along with its specific tools and methods, is significantly important in enabling the sustainable design of future machine components. It is urgent to create and enhance innovative solutions for upgrading the existing machine tools as well as for manufacturing processes and manufacturing systems. Therefore, several challenges lead to the future development of the industry of machine tools and their components (Garbie and Garbie, 2024).

The new sophisticated design of machines can be incorporated with sensor networking technology for the application of the Industrial Internet of Things (IIoT) and

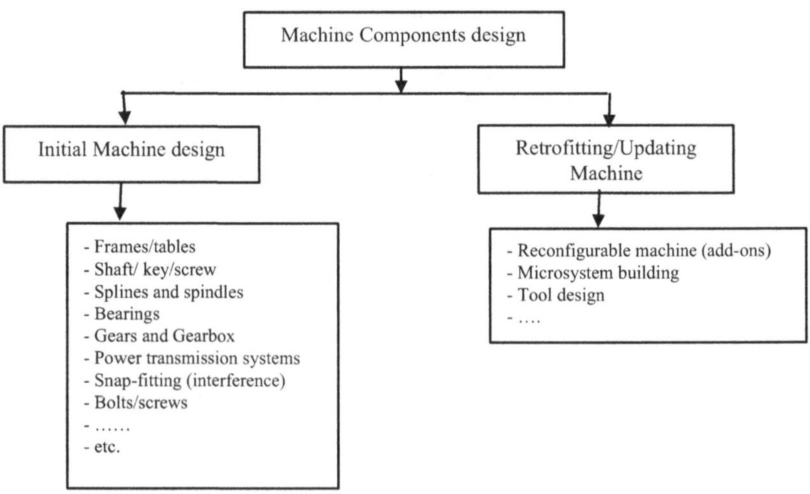

FIGURE 5.1 Classifications of machine components design.

wireless sensor nodes (WSN). These enablers are urgent for keeping a sustainable machine design, and they are so-called "microsystems technology". Reconfigurable machine tools are also considered one of the major enablers for maintaining sustainability in machine design. Reconfigurability allows for hardware upgrades to adapt to changes in market demand, such as increasing the capacity of the machine tool frame/structure, enhancing chucks to accept larger scales (diameter/length), improving moving and clamping devices and fixtures, modifying headstocks for greater scalability, introducing moving and rotating tables, implementing internally cooled tools, and embedding computer-integrated systems (e.g., CAD/CAM) (Garbie and Parsaei, 2022).

Machine tool design, initially, must be based on fastening more than joining to ensure ease of reconfiguration. The major difference between upgrading and retrofitting lies in the use of add-ons. Most add-ons are not permanent but flexible to meet required specifications. Therefore, smart manufacturing systems evolve by updating and upgrading according to the machine tool requirements for Industry 4.0 (Garbie and Garbie, 2022). Machine tool data are typically related to technologies such as the Industrial Internet of Things (IIoT), sensor networks, and cyber-physical systems. These data are analyzed through advanced data analytics, which is recommended and necessary for upgrading machine tools (Garbie and Garbie, 2020, 2022).

5.3 ANALYSIS AND ASSESSMENT OF SUSTAINABLE MACHINE/PRODUCT DESIGN (MD)

The main purpose of this section is to analyze and investigate the core layer of manufacturing, Machine Design (MD), to identify the enablers of sustainability by proposing mathematical models to estimate sustainability indexes. It can be noted from Figure 5.2 that the machine/product design layer (MD) of manufacturing

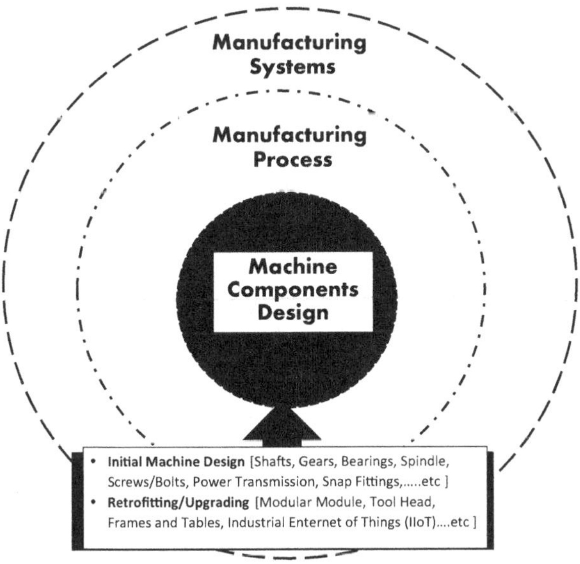

FIGURE 5.2 Machine components design concerning the whole manufacturing sector.

sustainability is separated into two fundamental parts: initial machine design and retrofitting/upgrading machine. Each has distinct characteristics in terms of components. In this context, machine/product design (MD) encompasses both the creation and optimization of mechanical components and the retrofitting of existing ones. These components include springs, shafts, and their components, gears, bearings, fasteners and joining mechanisms, and control transmission devices such as belts and flywheels. The concepts of MD address the design of these mechanical elements, as well as their retrofitting to enhance functionality and sustainability in alignment with evolving technological demands.

The sustainability enablers of the MD will include almost all aspects of the mechanical properties of counting tension (x_1), compression (x_2), impact (x_3), hardness (x_4), toughness (x_5), torsion (x_6), shear (x_7), corrosion (x_8), fatigue (x_9), fractures (x_{10}), creep (x_{11}), stiffness (x_{12}) and bending and flexure (x_{13}), modular models (x_{14}), device head (x_{15}), frames, moving and rotating tables (x_{16}), the Industrial Internet of Things (x_{17}), wireless sensor systems (x_{18}), cyber-physical systems (x_{19}), embedded computer frameworks (x_{20}), and information analytics (x_{21}),...x_n, etc. (Table 5.1).

TABLE 5.1
Sustainability Enablers of Machine/Product Design

Machine Design (MD)		Enablers of Components/Parts	Sustainability Pillars Concentration
Design of machines "initial"	Shafts Gears Bearings Spindles Screws/bolts Transmission Snap-fitting	• Tension (x_1) • Compression (x_2) • Impact (x_3) • Hardness (x_4) • Toughness (x_5) • Torsion (x_6) • Shear (x_7) • Corrosion (x_8) • Fatigue (x_9) • Fractures (x_{10}) • Creep (x_{11}) • Stiffness (x_{12}) • Bending and Flexure (x_{13})	Economy and environment
Retrofitting/ Up-grading machine	Reconfigurable parts (RM)	• Modular modules (x_{14}) • Device head changing (x_{15}) • Adjustable components (x_{16})	Economy and environment
	Microsystem technology (MT)	• Industrial internet of things (x_{17}) • Wireless sensor networks, WSN (x_{18}) • Cyber-physical systems, CPS (x_{19}) • Embedded computed integrated systems (e.g., CAD/CAM) (ECS) (x_{20}) • Information analytics (x_{21})	

The sustainability index of the MD at any time t, $S_{MD}(t)$, can be numerically communicated as a work of all past enablers or components as appeared within the following Equation (5.1). This can be a so-called center condition of manufacturing sustainability.

$$S_{MD}(t) = f(x_1, x_2, x_3, \ldots x_n) = \left\{ \begin{array}{c} x_1 \\ x_2 \\ x_3 \\ \text{—} \\ x_n \end{array} \right\} = \sum_{i=1}^{n} x_i(t)/n \tag{5.1}$$

where:

$x_i(t)$ in Equation (5.1) can be calculated using the following Equation (5.2)

$$x_i(t) = \frac{x_i(t)\text{current} - x_i(t)\min}{x_i(t)\max - x_i(t)\min} \tag{5.2}$$

where:

$x_i(t)$ is the normalized value of the enabler of the property specified at time t,

$x_i(t)$current is the existing value of the property specified in any part of the machine at any time t,

$x_i(t)$max and $x_i(t)$min are values of the property specified at time t, respectively.

n is the number of aspects of the MD.

$x_1(t)$=component tension level at any time t.

$$x_1(t) = \frac{E^{x1} - I^{x1}_{\min}}{I^{x1}_{\max} - I^{x1}_{\min}} \tag{5.3}$$

E^{x1} = the existing value of component tension at time t,

$I^{x1}_{\max}$ = maximum value of component tension at time t,

$I^{x1}_{\min}$ = minimum value of component tension at time t,

$x_2(t)$ = component compression level at any time t.

$$x_2(t) = \frac{E^{x2} - I^{x2}_{\min}}{I^{x2}_{\max} - I^{x2}_{\min}} \tag{5.4}$$

E^{x2} = the existing value of component compression at time t,

$I^{x2}_{\max}$ = maximum value of component compression at time t,

$I^{x2}_{\min}$ = minimum value of component compression at time t,

$x_3(t)$=component impact level at any time t.

$$x_3(t) = \frac{E^{x3} - I^{x3}_{\min}}{I^{x3}_{\max} - I^{x3}_{\min}} \tag{5.5}$$

E^{x3} = the existing value of component impact at time t,

I_{max}^{x3} = maximum value of component impact at time t,
I_{min}^{x3} = minimum value of component impact at time t,
$x_4(t)$ = component hardness level at any time t.

$$x_4(t) = \frac{E^{x4} - I_{min}^{x4}}{I_{max}^{x4} - I_{min}^{x4}} \tag{5.6}$$

E^{x4} = the existing value of component hardness at time t,
I_{max}^{x4} = maximum value of component hardness at time t,
I_{min}^{x4} = minimum value of component hardness at time t,
$x_5(t)$ = component toughness level at any time t.

$$x_5(t) = \frac{E^{x5} - I_{min}^{x5}}{I_{max}^{x5} - I_{min}^{x5}} \tag{5.7}$$

E^{x5} = the existing value of component toughness at time t,
I_{max}^{x5} = maximum value of component toughness at time t,
I_{min}^{x5} = minimum value of component toughness at time t,
$x_6(t)$ = component torsion level at any time t.

$$x_6(t) = \frac{E^{x6} - I_{min}^{x6}}{I_{max}^{x6} - I_{min}^{x6}} \tag{5.8}$$

E^{x6} = the existing value of component torsion at time t,
I_{max}^{x6} = maximum value of component torsion at time t,
I_{min}^{x6} = minimum value of component torsion at time t,
$x_7(t)$ = component shear level at any time t.

$$x_7(t) = \frac{E^{x7} - I_{min}^{x7}}{I_{max}^{x7} - I_{min}^{x7}} \tag{5.9}$$

E^{x7} = the existing value of component shear at time t,
I_{max}^{x7} = maximum value of component shear at time t,
I_{min}^{x7} = minimum value of component shear at time t,
$x_8(t)$ = component corrosion at time t.

$$x_8(t) = \frac{E^{x8} - I_{min}^{x8}}{I_{max}^{x8} - I_{min}^{x8}} \tag{5.10}$$

E^{x8} = the existing value of component corrosion at time t,
I_{max}^{x8} = maximum value of component corrosion at time t,
I_{min}^{x8} = minimum value of component corrosion at time t,
$x_9(t)$ = component fatigue at time t.

$$x_9(t) = \frac{E^{x9} - I_{min}^{x9}}{I_{max}^{x9} - I_{min}^{x9}} \tag{5.11}$$

E^{x9} = the existing value of component fatigue at time t,
I_{max}^{x9} = maximum value of component fatigue at time t,
I_{min}^{x9} = minimum value of component fatigue at time t,
$x_{10}(t)$ = component fractures level at any time t.

$$x_{10}(t) = \frac{E^{x10} - I_{min}^{x10}}{I_{max}^{x10} - I_{min}^{x10}} \tag{5.12}$$

E^{x10} = the existing value of component fractures level at any time t,
I_{max}^{x10} = maximum value of component fractures level at any time t,
I_{min}^{x10} = minimum value of component fractures level at any time t,
$x_{11}(t)$ = component creep level at any time t.

$$x_{11}(t) = \frac{E^{x11} - I_{min}^{x11}}{I_{max}^{x11} - I_{min}^{x11}} \tag{5.13}$$

E^{x11} = the existing value of component creep at any time t,
I_{max}^{x11} = maximum value of component creep at any time t,
I_{min}^{x11} = minimum value of component creep at any time t,
$x_{12}(t)$ = component stiffness level at any time t.

$$x_{12}(t) = \frac{E^{x12} - I_{min}^{x12}}{I_{max}^{x12} - I_{min}^{x12}} \tag{5.14}$$

E^{x12} = the existing value of component stiffness at any time t,
I_{max}^{x12} = maximum value of component stiffness at any time t,
I_{min}^{x12} = minimum value of component stiffness at any time t,
$x_{13}(t)$ = component bending and flexure at time t.

$$x_{13}(t) = \frac{E^{x13} - I_{min}^{x13}}{I_{max}^{x13} - I_{min}^{x13}} \tag{5.15}$$

E^{x13} = the existing value of component bending and flexure at time t,
I_{max}^{x13} = maximum value of component bending and flexure at time t,
I_{min}^{x13} = minimum value of component bending and flexure at time t,
$x_{14}(t)$ = normalizing modular modules at time t.

$$x_{14}(t) = \frac{E^{x14} - I_{min}^{x14}}{I_{max}^{x14} - I_{min}^{x14}} \tag{5.16}$$

E^{x14} = the existing number of modular modules at time t,
I_{max}^{x14} = maximum number of modular modules at time t,
I_{min}^{x14} = minimum number of modular modules at time t,
$x_{15}(t)$ = normalizing device head level at time t.

$$x_{15}(t) = \frac{E^{x15} - I_{min}^{x15}}{I_{max}^{x15} - I_{min}^{x15}} \tag{5.17}$$

E^{x15} = the existing (%) of using device head at time t,

I_{max}^{x15} = maximum (%) of using device head at time t,

I_{min}^{x15} = minimum (%) of using device head at time t,

$x_{16}(t)$ = normalizing frames, moving and rotating tables level at time t.

$$x_{16}(t) = \frac{E^{x16} - I_{min}^{x16}}{I_{max}^{x16} - I_{min}^{x16}} \tag{5.18}$$

E^{x16} = the existing (%) of using frames, moving and rotating tables at time t,

I_{max}^{x16} = maximum (%) of using frames, moving and rotating tables at time t,

I_{min}^{x16} = minimum (%) of using frames, moving and rotating tables at time t.

$x_{17}(t)$ = normalizing using the IoT level at time t.

$$x_{17}(t) = \frac{E^{x17} - I_{min}^{x17}}{I_{max}^{x17} - I_{min}^{x17}} \tag{5.19}$$

E^{x17} = the existing (%) of using the IoT level at time t,

I_{max}^{x17} = maximum (%) of using the IoT level at time t,

I_{min}^{x17} = minimum (%) of using the IoT level at time t,

$x_{18}(t)$ = normalizing using the wireless sensor networks level at time t.

$$x_{18}(t) = \frac{E^{x18} - I_{min}^{x18}}{I_{max}^{x18} - I_{min}^{x18}} \tag{5.20}$$

E^{x18} = the existing (%) of using the wireless sensor networks level at time t,

I_{max}^{x18} = maximum (%) of using the wireless sensor networks level at time t,

I_{min}^{x18} = minimum (%) of using the wireless sensor networks level at time t,

$x_{19}(t)$ = normalizing using the cyber physical production systems (CPPS) level at time t.

$$x_{19}(t) = \frac{E^{x19} - I_{min}^{x19}}{I_{max}^{x19} - I_{min}^{x19}} \tag{5.21}$$

E^{x19} = the existing (%) of using the CPPS level at time t,

I_{max}^{x19} = maximum (%) of using the CPPS level at time t,

I_{min}^{x19} = minimum (%) of using the CPPS level at time t,

$x_{20}(t)$ = normalizing using embedded integrated systems level at time t.

$$x_{20}(t) = \frac{E^{x20} - I_{min}^{x20}}{I_{max}^{x20} - I_{min}^{x20}} \tag{5.22}$$

E^{x20} = the existing percentage of using embedded integrated level at time t,

I_{max}^{x20} = maximum percentage of using embedded integrated level at time t,

I_{min}^{x20} = minimum percentage of using embedded integrated level at time t,

$x_{21}(t)$ = normalizing using information analytics systems level at time t.

$$x_{21}(t) = \frac{E^{x21} - I_{min}^{x21}}{I_{max}^{x21} - I_{min}^{x21}} \tag{5.23}$$

E^{x21} = the existing percentage of using information analytics systems level at time t,
$I_{\max}^{x21}$ = maximum percentage of using information analytics systems level at time t,
$I_{\min}^{x21}$ = minimum percentage of using information analytics systems level at time t.

5.4 NUMERICAL EXAMPLE 5.1

This numerical illustration demonstrates the proposed method or system for evaluating the sustainable manufacturing index in terms of machine design. It achieves this by distinguishing sustainability enablers and performance measurements, including existing values and target values (minimum and maximum). Relevant information regarding sustainability enablers is presented in Table 5.2 for machine design.

Concerning the sustainability of machine design (Table 5.2), the foremost imperative issue is normalizing the enablers of sustainability through their indicators. For example, the indicator for tensile quality (x_1), which is evaluated as the tensile strength in any component of the machine at any time by utilizing two values for the target (greatest as 100 MPa and least as 85 MPa) and one existing value as 95 MPa. At that point, the normalized value of tensile strength is assessed as 0.50 as given in the following equation and appears in Table 5.3. The total assessed indicators result in machine design (MD) sustainability, which appears in Table 5.3.

$$x_1(t) = \frac{95-85}{100-85} = 0.67$$

TABLE 5.2
Data for Machine/Product Components Design

Indicator (x)	Units Performance Measures	Performance metrics: Existing Value	Target Value: Max	Target Value: Min
x_1	Value of tensile strength at time t, MPa	95	100	85
x_2	Value of compression at time t, MPa	100	110	95
x_3	Value of impact at time t, MPa	70	80	60
x_4	Value of hardness at time t, MPa	195	200	190
x_5	Value of toughness at time t, MPa	145	150	140
x_6	Value of torsion at time t, MPa	143	145	140
x_7	Value of shear stress at time t, MPa	170	172	168
x_8	Value of corrosion at time t, mm/year	0.009	0.01	0.007
x_9	Value of fatigue at time t, MPa	210	220	200

(Continued)

TABLE 5.2 (*Continued*)
Data for Machine/Product Components Design

Indicator (x)	Units Performance Measures	Performance metrics: Existing Value	Target Value: Max	Target Value: Min
x_{10}	Value of fracture at time t, MPa	450	500	400
x_{11}	Value of creep at time t, mm/year	0.008	0.01	0.006
x_{12}	Value of stiffness at time t, MPa	95	100	90
x_{13}	Value of bending and flexure at time t, MPa	195	200	190
x_{14}	Number of modular modules in the machine at any time t	8	10	6
x_{15}	Possibility to change the head of a machine at time t	90	100	80
x_{16}	Number of frames, moving and rotating tables in a machine at any time t	3	4	2
x_{17}	Availability of using the industrial internet of thing (IIOT) at any time t	90	95	85
x_{18}	Number of sensors at the time t	10	12	8
x_{19}	Percentage of integration of CPS at time t	85	90	80
x_{20}	Availability of using embedded computer systems (ECS) at time t	85	95	80
x_{21}	Availability of using (DA) at time t	80	90	70

The sustainability index of machine components designs at any time t, $s_{\text{MD}}(t)$, is evaluated utilizing Equation (5.1), and the results are outlined in Table 5.3 as 0.4814 (48.14%). The sustainable index of machine design currently stands at 48.14%, which is relatively low and requires improvement across all aspects of machine design, including initial design components, reconfigurability, and microsystem technology.

5.5 CONCLUDING REMARKS

In this chapter, the first layer of manufacturing sustainability based on the architecture system perspective was proposed and analyzed through the core layer which was known as machine/product components design. The concepts of the core layer, which was based on machine/product components design were analyzed and presented within the manufacturing world as a novel and inventive approach. In the machine/product components design, the essential enablers were recognized, and the sustainability index was assessed. It was logically observed that the core layer of sustainable manufacturing encompasses several critical enablers necessary for achieving sustainability. The architecture system plays a pivotal role in sustainable manufacturing by linking the core layer to the subsequent layers referred to as the "chosen layer" and the "given layer", as illustrated in Figure 5.3.

TABLE 5.3
Sustainability Index in Machine/Product Design

Indicator (x)	x_1	x_2	x_3	x_4	x_5	x_6	x_7	x_8	x_9	x_{10}	x_{11}	x_{12}	x_{13}	x_{14}	x_{15}	x_{16}	x_{17}	x_{18}	x_{19}	x_{20}	x_{21}
Normalized Index	0.66	0.33	0.5	0.5	0.5	0.67	0.5	0.66	0.5	0.5	0.5	0.5	0.5	0.5	0.5	0.5	0.5	0.5	0.5	0.33	0.5
$S_{MD}(t)$	0.4814																				

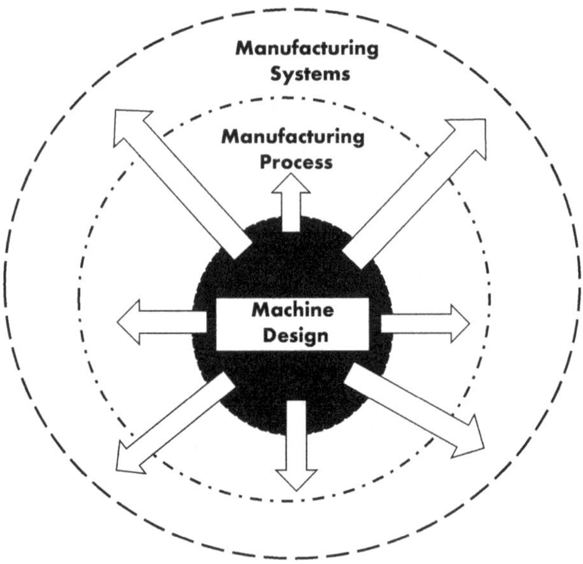

FIGURE 5.3 Affecting core layer (machine design) on the chosen layer (manufacturing processes) and given layer (manufacturing systems).

REFERENCES

Garbie, I., (2014), "An Analytical Technique to Model and Assess Sustainable Development Index in Manufacturing Enterprises", *International Journal of Production Research*, Vol. 52, No. 16, pp. 4876–4915.

Garbie, I., and Garbie, A., (2020), "Sustainability and Manufacturing: A Conceptual Approach", *Proceedings of the Industrial and Systems Engineering Research Conference (IISE Annual Conference and Expo 2020)*, November 1–3, 2020, New Orleans, LA, USA, 6 pp, Held Virtually.

Garbie, I., and Garbie, A., (2022), "A Holistic Perspective of Sustainable Manufacturing: A Novel Conceptual Approach", *International Journal of Industrial and Systems Engineering*, Vol. 41, No. 2, pp. 135–167.

Garbie, I., and Parsaei, H., (2022), *Reconfigurable Manufacturing Enterprises for Industry 4.0*, 1st Edition. CRC Press-Taylor and Francis, Boca Raton, FL, USA.

Garbie, I., and Garbie, A., (2024a), "Toward Smart Manufacturing Systems incorporating Reconfiguration Issues", *International Journal of Industrial and Systems Engineering*, Vol. 46, No. 1, pp. 1–33, 2024.

Garbie, I., and Garbie, M., (2024b), "Sustainability of Manufacturing Enterprises: A Psychological Perspective of Linking Manufacturing Layers", *Proceedings of the Industrial and Systems Engineering Research Conference (IISE Annual Conference and Expo 2024)*, May 18–21, 2024, Montreal, Canada, 6 pp.

Kianinejad, K., Thorn, S., Kushwaha, S., and Uhlmann, E., (2018), "Add-on Error Compensation Unit as Sustainable Solution for Outdated Milling Machines", *Procedia CIRP*, Vol. 40, pp. 174–178.

Shneor, Y., (2018), "Reconfigurable Machine Tool: CNC Machine for Milling Grinding and Polishing", *Procedia Manufacturing*, Vol. 21, pp. 221–227.

6 Sustainable Manufacturing Processes

6.1 INTRODUCTION TO MANUFACTURING PROCESSES (MP)

Manufacturing processes encompass systematic procedures and technologies used to transform materials from their initial state, from raw material or semi-finished products, into a final product through a specific sequence of operations. These processes are broadly divided into two primary categories: processing operations and assembly operations (Figure 6.1). For each unique operation, different sub-operations are created (Garbie and Garbie, 2022; Garbie and Parsaei, 2022).

The manufacturing processes (metal forming, machining, etc.) still require further work in terms of establishing a sophisticated approach or algorithm to analyze the sustainability level opposite the manufacturing systems (Garbie, 2014; Garbie and Garbie, 2020). The importance of economic and environmental aspects was mentioned in manufacturing processes (Jayakrishna et al., 2016). Also, the manufacturing

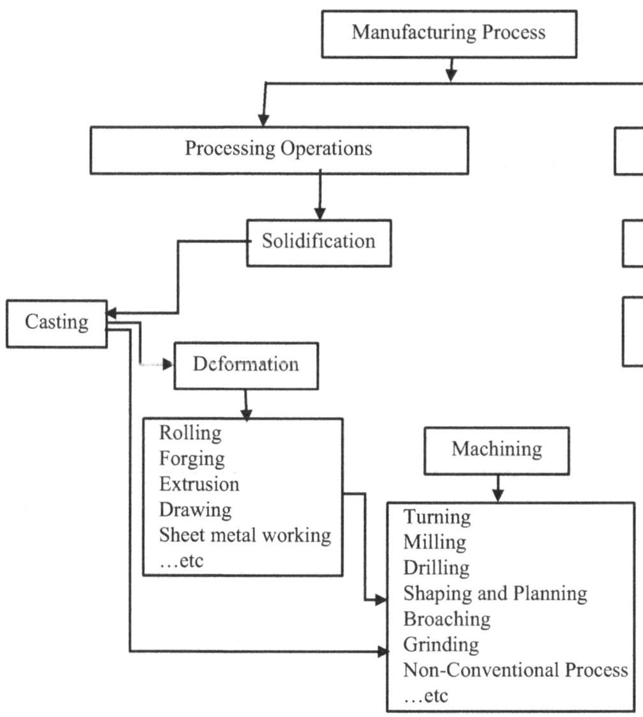

FIGURE 6.1 Classifications of the manufacturing process.

DOI: 10.1201/9781003644835-9

processes need a design and management of social and environmental requirements with economic requirements. Therefore, it is urgent to formulate a general mathematical model to assess or estimate the sustainability of manufacturing as a whole and manufacturing processes specifically. Affecting the hazardous cutting fluids during the machining operations is still considered an urgent issue with emphasizing on the consumption of energy and natural resources (Goindi and Sarkar, 2017; Garbie and Garbie, 2022). Environmental effects are discussed and mentioned as one of the major issues in manufacturing processes through using cryogenic machining (Lu et al., 2018a,b).

Although manufacturing processes include several operations, the machining process is one of the major processes that is carried out to produce the final shape of the part. All of these processes are energy-intensive. Energy is one of these enablers of sustainability in manufacturing processes. Manufacturing sustainability assessment of the machining process is considered an essential requirement for providing optimal cutting conditions (Hegab et al., 2018). Moreover, the whole manufacturing sustainability includes sustainable manufacturing processes. The machining process of "metal removal" represents the main manufacturing activity deserving more focus and concentration because of its different enablers and variety. Also, dry flood coolants and water vapor are used to investigate sustainability assessment in manufacturing processes through the quality of the surface due to damage to the surface and the generating residual stresses (Kadam and Pawade, 2017). Therefore, the most important question related to sustainable manufacturing processes is where the enablers in productive processes regarding manufacturing sustainability (Garbie and Garbie, 2024a,b).

6.2 MANUFACTURING PROCESSES WITHIN AN ARCHITECTURE PERSPECTIVE

It seems that the concept of the manufacturing processes is based on classifying the four common main manufacturing operations as either processing or assembly such as solidification (e.g., casting), deformation (e.g. forging, rolling, extrusions, sheet metal working), machining (turning, milling, drilling, shaping, grinding), assembly (permanent joining and mechanical fastening), etc. (Figure 6.2).

Sustainable manufacturing processes rely on the mentioned productive operations in the previous paragraph. Enablers analysis is illustrated in Table 6.1. Table 6.2 identifies the sustainability enablers with their productive processes.

According to Figure 6.2 and Table 6.2, the manufacturing process (MP) is based on the core layer "machine/product components design" and is partitioned into four major operations beginning with solidification, deformation, machining, and finally assembly. For illustration, a few enablers must exist to keep the sustainability of solidification, for example, foundry or die, energy reserved, die design, and die life. Regarding deformation, the energy saved, and die design are the foremost critical enablers in addition to rolling, forging, extrusion, drawing, and sheet metalworking for bulk or sheet metalworking. Machining processes represent basically energy required, tool geometry or tool design, tool life, and cooling systems (coolant) with all diverse types of machining operations: turning, milling, drilling, shaping, and

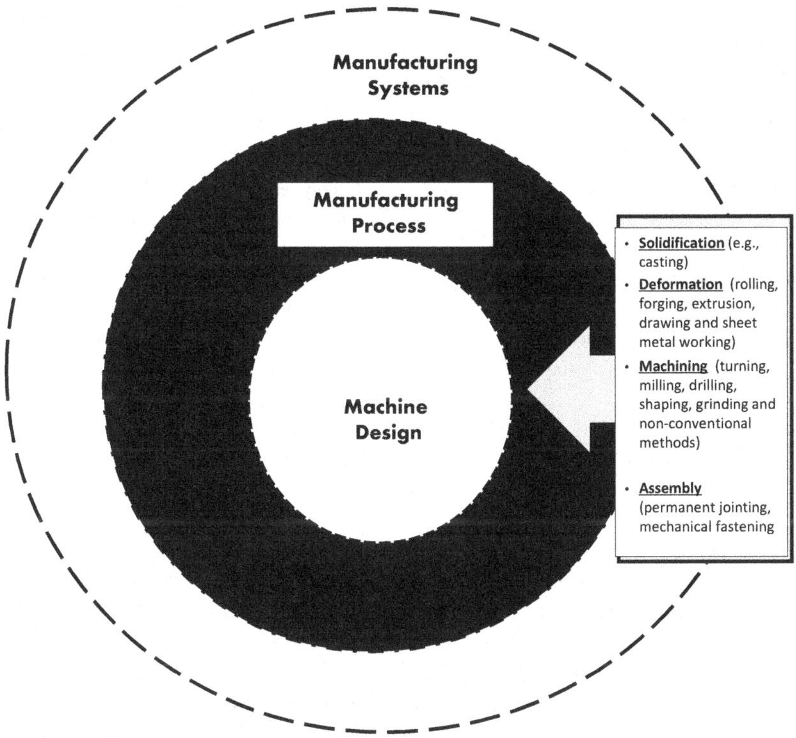

FIGURE 6.2 Manufacturing processes concerning the whole manufacturing.

grinding. Concerning assembly, energy reserved and easy to assemble/disassemble are represented as the enablers of assembly operations (Garbie and Garbie, 2022).

6.3 ANALYSIS AND ASSESSMENT OF SUSTAINABLE MANUFACTURING PROCESSES

Although the sustainability of manufacturing processes (MP) depends on these enablers "energy saved (y_1), die life (y_2), die design (y_3), tool geometry (y_4), tool life (y_5), cooling systems (y_6), and easy to assembly/disassembly (y_7), y_m, etc.", their operations and machines are still characterized as said in the core layer "foundry, rolling, forging, extrusion, drawing, turning, milling, drilling, shaping, and grinding". At that point, the sustainability of manufacturing processes (MP), $S_{MP}(t)$, numerically defined as the equation as a function (6.1) of the past enablers.

$$S_{MP}(t) = f(y_1, y_2, y_3, \dots y_n) = \left\{ \begin{array}{c} y_1 \\ y_2 \\ y_3 \\ \text{---} \\ y_m \end{array} \right\} = \sum_{i=1}^{m} y_i(t)/m \tag{6.1}$$

TABLE 6.1
Enablers of Manufacturing Processes

Manufacturing Process Types (MP)		Enablers/Incentives/Drivers/Motivation	Sustainability Pillars Concentration
Solidification, "e.g., casting"		Energy reserved, y_1 Die life, y_2 Die design, y_3	Economy and environment
Bulk metal deformation	Rolling Forging Extrusion Drawing	Energy reserved, y_1 Die design, y_3	Economy and environment
Sheet metalworking		Energy reserved, y_1 Die design, y_3	Economy
Metal machining	Turning Milling Drilling Shaping Broaching Grinding Non-conventional machining	Energy reserved, y_1 Tool geometry, y_4 Tool life, y_5 Cooling system (coolant), y_6	Economy, society, and environment
Assembly operations	Permanent joint "welding" Permanent joint "riveting" Fastening "bolts" Screws "screws"	Energy reserved, y_1 Easy to assembly/disassembly, y_7	Economy, society, and environment

where:

$y_i(t)$ in Equation (6.1) is calculated using the following Equation (6.2)

$$y_i(t) = \frac{y_i(t)\text{current} - y_i(t)\text{min}}{y_i(t)\text{max} - y_i(t)\text{min}} \tag{6.2}$$

where:

$y_i(t)$ is the normalized value of the enabler with the manufacturing process (MP) at any time t,

$y_i(t)$ current is the existing value of the enabler with the manufacturing process at any time t,

$y_i(t)\text{max}$ and $y_i(t)\text{min}$ are the maximum and minimum values of the enabler with the manufacturing process at any time t, respectively,

m is the number of enablers of the MP,

$y_1(t)$ = normalizing energy reserved level at any time t.

$$y_1(t) = \frac{E^{y_1} - I_{\text{min}}^{y_1}}{I_{\text{max}}^{y_1} - I_{\text{min}}^{y_1}} \tag{6.3}$$

E^{y_1} = the existing value of energy consumed at any time t,

TABLE 6.2
Data for Manufacturing Processes

Indicator (y)	Units Performance Measures	Existing Value	Target Value Max	Target Value Min
y_1	Usage of energy in casting/solidification at any time t, kw	1500	2000	1000
y_2	Number of products manufactured during the die life at any time t	120 K	150 K	100 K
y_3	Availability of designing a new die at any time t for bulk deformation	90	95	80
y_4	Availability of configuration and cutting angles of a cutting tool at any time t	95	100	90
y_5	Cutting/machining time before replacing the cutting tool at any time t, minutes	15	25	10
y_6	Availability of cooling (coolant) type at any time t	90	95	85
y_7	Degree of difficulty to assembly/disassembly at any time t	18	20	15

(Existing Value, Max and Min columns fall under "Performance Metrics".)

$I_{max}^{y_1}$ = maximum value of energy consumed at any time t,
$I_{min}^{y_1}$ = minimum value of energy consumed at any time t,
$y_2(t)$ = normalizing die life level at any time t.

$$y_2(t) = \frac{E^{y_2} - I_{min}^{y_2}}{I_{max}^{y_2} - I_{min}^{y_2}} \tag{6.4}$$

E^{y_2} = the existing products manufactured during the die life at time t,
$I_{max}^{y_2}$ = maximum products manufactured during the die life at time t,
$I_{min}^{y_2}$ = minimum products manufactured during the die life at time t,
$y_3(t)$ = normalizing die design level at any time t.

$$y_3(t) = \frac{E^{y_3} - I_{min}^{y_3}}{I_{max}^{y_3} - I_{min}^{y_3}} \tag{6.5}$$

E^{y_3} = the existing availability of designing a new die at any time t,
$I_{max}^{y_3}$ = maximum availability of designing a new die at any time t,
$I_{min}^{y_3}$ = minimum availability of designing a new die at any time t,
$y_4(t)$ = normalizing tool geometry level at any time t.

$$y_4(t) = \frac{E^{y_4} - I_{min}^{y_4}}{I_{max}^{y_4} - I_{min}^{y_4}} \tag{6.6}$$

E^{y_4} = the existing availability of configuration and cutting angles of a cutting tool at any time t,

$I_{max}^{y_4}$ = maximum availability of configuration and cutting angles of a cutting tool at any time t,

$I_{\min}^{y4}$ = minimum availability of configuration and cutting angles of a cutting tool at any time t,

$y_5(t)$ = normalizing tool life level at any time t.

$$y_5(t) = \frac{E^{y5} - I_{\min}^{y5}}{I_{\max}^{y5} - I_{m5n}^{y5}} \tag{6.7}$$

E^{y5} = the existing cutting/machining time before replacing the cutting tool at any time t,

$I_{\max}^{y5}$ = maximum cutting/machining time before replacing the cutting tool at any time t,

$I_{\min}^{y5}$ = minimum cutting/machining time before replacing the cutting tool at any time t,

$y_6(t)$ = normalizing cooling systems level at any time t.

$$y_6(t) = \frac{E^{y6} - I_{\min}^{y6}}{I_{\max}^{y6} - I_{m5n}^{y6}} \tag{6.8}$$

E^{y6} = the existing availability of cooling (coolant) type at any time t,

$I_{\max}^{y6}$ = maximum availability of cooling (coolant) type at any time t,

$I_{\min}^{y6}$ = minimum availability of cooling (coolant) type at any time t,

$y_7(t)$ = normalizing easy to assembly/disassembly level at any time t.

$$y_7(t) = \frac{E^{y7} - I_{\min}^{y7}}{I_{\max}^{y7} - I_{m5n}^{y7}} \tag{6.9}$$

E^{y9} = the existing degree of difficulty to assembly/disassembly at any time t,

$I_{\max}^{y7}$ = maximum degree of difficulty to assembly/disassembly at any time t,

$I_{\min}^{y7}$ = minimum degree of difficulty to assembly/disassembly at any time t.

6.4 NUMERICAL EXAMPLE 6.1

This numerical example demonstrates the proposed method for evaluating the sustainable manufacturing index concerning manufacturing processes. It does so by identifying sustainability enablers and performance measurements, including existing values and target values (minimum and maximum). Relevant information about these sustainability enablers is presented in Table 6.2 for manufacturing processes.

Concerning the sustainability of manufacturing processes (Table 6.2), the foremost imperative issue is normalizing the enablers of sustainability through their indicators. For example, the indicator for casting/solidification (y_1), which is evaluated as the usage of energy in any part produced of the manufacturing at any time by utilizing two values for the target (greatest as 2000 kW and least as 1000 kW) and one existing value as 1500 kW. At that point, the normalized value of casting/solidification is assessed as 0.50 as the following equation and appears in Table 6.3. The sustainability index of manufacturing processes (MP) is also displayed in Table 6.3.

$$y_1(t) = \frac{1500 - 1000}{2000 - 1000} = 0.50$$

TABLE 6.3
Sustainability Index in Manufacturing Processes

Indicator (y)	y_1	y_2	y_3	y_4	y_5	y_6	y_7
Normalized index	0.50	0.40	0.67	0.50	0.30	0.50	0.60
$S_{MP}(t)$	0.4957						

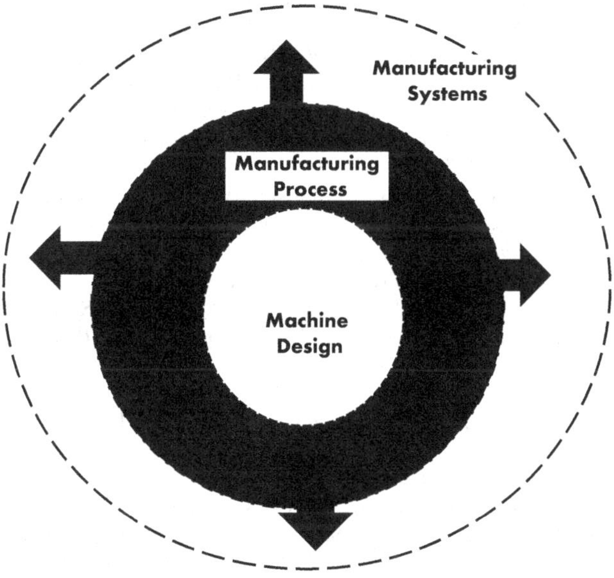

FIGURE 6.3 Affecting chosen layer (manufacturing processes) on the given layer (manufacturing systems).

The sustainability index of manufacturing processes at any time t, $s_{MD}(t)$, is evaluated utilizing Equation (6.1), and the results are outlined in Table 6.3 as 0.4957 (49.57%).

6.5 CONCLUDING REMARKS

The second layer of manufacturing sustainability, viewed from an architectural perspective, was proposed and analyzed through the chosen layer, identified as manufacturing processes. This layer, based on the four major manufacturing processes—processing operations, and assembly operations—was examined and presented as a novel and innovative approach within the manufacturing domain.

In the context of manufacturing processes, common essential enablers were identified, and the sustainability index was evaluated. It was logically observed that the chosen layer of sustainable manufacturing encompasses several key enablers required to achieve sustainability. The architectural system plays a pivotal role in sustainable manufacturing by linking the chosen layer to the subsequent "given layer", as illustrated in Figure 6.3.

REFERENCES

Garbie, I., (2014), "An Analytical Technique to Model and Assess Sustainable Development Index in Manufacturing Enterprises," *International Journal of Production Research*, Vol. 52, No. 16, pp. 4876–4915.

Garbie, I., and Garbie, A., (2020), "Sustainability and Manufacturing: A Conceptual Approach", *Proceedings of the Industrial and Systems Engineering Research Conference (IISE Annual Conference and Expo 2020)*, November 1–3, 2020, New Orleans, LA, USA, 6 pp, Held Virtually.

Garbie, I., and Garbie, A., (2022), "A Holistic Perspective of Sustainable Manufacturing: A Novel Conceptual Approach", *International Journal of Industrial and Systems Engineering*, Vol. 41, No. 2, pp. 135–167.

Garbie, I., and Parsaei, H., (2022), *Reconfigurable Manufacturing Enterprises for Industry 4.0*, 1st Edition. CRC Press-Taylor & Francis, Boca Raton, FL, USA.

Garbie, I., and Garbie, A., (2024a), "Toward Smart Manufacturing Systems incorporating Reconfiguration Issues", *International Journal of Industrial and Systems Engineering*, Vol. 46, No. 1, pp. 1–33, 2024.

Garbie, I., and Garbie, M., (2024b), "Sustainability of Manufacturing Enterprises: As Psychological Perspective of Linking Manufacturing Layers", *Proceedings of the Industrial and Systems Engineering Research Conference (IISE Annual Conference and Expo 2024)*, May 18–21, 2024, Montreal, Canada, 6 pp.

Goindi, G. S., and Sarkar, P., (2017), "Dry Machining: A Step towards Sustainable Machining-Challenges and Future Directions", *Journal of Cleaner Production*, Vol. 165, pp. 1557–1571.

Hegab, H. A., Darras, B., and Kishawy, H. A., (2018), "Towards Sustainability Assessment of Machining Processes", *Journal of Cleaner Production*, Vol. 170, pp. 694–703.

Jayakrishna, A., Girubha, R. J., and Vindho, S., (2016), "Comparison of Sustainability Characteristics of Conventional and CNC Turning Processes, A Case Study", *Journal of Engineering Design and Technology*, Vol. 14, No. 3, pp. 422–445.

Kadam, G. S., and Pawade, R. S., (2017), "Surface Integrity and Sustainability Assessment in High-Speed Machining of Inconel 718: An Eco-Friendly Green approach", *Journal of Cleaner Production*, Vol. 147, pp. 273–283.

Lu, T., Kudaravalli, R., and Georgiou, G., (2018a), "Cryogenic Machining through the Spindle and Tool for Improved Machining Process Performance and Sustainability: Pt. I, Systems Design", *Procedia Manufacturing*, Vol. 21, pp. 266–272.

Lu, T., Kudaravalli, R., and Georgiou, G., (2018b), "Cryogenic Machining through the Spindle and Tool for Improved Machining Process Performance and Sustainability: Pt. II, Sustainability Performance Study", *Procedia Manufacturing*, Vol. 21, pp. 273–280.

7 Sustainable Manufacturing Systems

7.1 INTRODUCTION TO MANUFACTURING SYSTEMS (MS)

As manufacturing systems consist of humans, resources (pieces of equipment, material handling, machines tools, etc.), and those procedures to implement the manufacturing processes, they can be the so-called design and management of manufacturing processes and their associated machines. Designing and installing different types of manufacturing systems according to the nature of products is one of the major challenges facing manufacturing systems designers and industrialists today. The three common types of manufacturing systems are: functional or process layout "job shop", product layout "flow shop", and focused layout "cellular". Although designing functional or process layout and product layout systems does not rely on a scientific technique and follow up the sequence of operations, respectively, it is recommended to establish them at least as functional or reminder and dedicated systems in designing hybrid manufacturing systems. Focused layout "cellular" systems have many approaches and methodologies to cluster machines into machine cells and group components parts into part families taking into consideration so many issues in terms of productivity and flexibility. For these reasons, focused layout systems are most recommended for hybrid manufacturing systems as the core of installing this type of facility design to overcome those deficiencies coming from the previous ones (Garbie and Parsaei, 2022; Garbie and Garbie, 2024a,b,c).

Designing the facilities layout for sustainability has several issues more than traditional methods (Garbie, 2013, 2014) such as globalization and international, contemporary, innovation, reconfiguration, lean, agile, flexibility in evaluation, and liquidity in management (Figure 7.1). Each issue is identified by enablers that make it sustainable. Therefore, the question related to facilities design and layout systems is where the enablers in manufacturing sustainability are (Garbie and Garbie, 2022).

It can be noticed from the analysis of manufacturing systems that mainly rely on manufacturing strategies and philosophies, not specific techniques or approaches as mentioned in previous analyses of either machine design or manufacturing processes. Garbie (2013, 2014) and Garbie and Garbie (2022) conducted a comprehensive survey regarding manufacturing sustainability in facilities design and layout systems. The main objective of this chapter is to formulate and model the sustainability index for the given layer of manufacturing as a single unit.

DOI: 10.1201/9781003644835-10

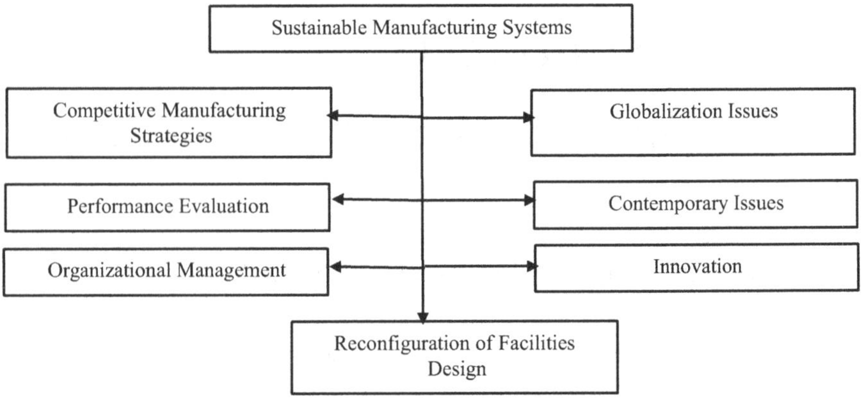

FIGURE 7.1 Different issues for keeping manufacturing systems sustainable.

7.2 MANUFACTURING SYSTEMS WITHIN AN ARCHITECTURE PERSPECTIVE

Manufacturing systems from the architecture perspective represent the last layer of sustainable manufacturing as shown in Figure 7.2. It seems from Figure 7.2 that the concept of the manufacturing systems is based on classifying four main manufacturing systems design types: functional or process layout "job shop", product layout "flow shop", cellular (focused) manufacturing system, and finally the assembly system.

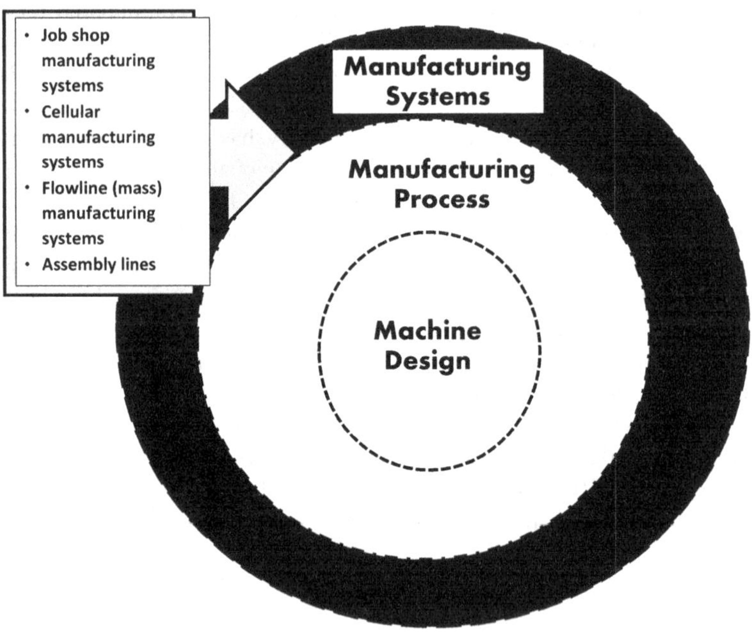

FIGURE 7.2 Manufacturing Systems concerning the whole manufacturing.

The sustainability of facilities design and layout systems mainly relies on major enablers which are listed and summarized in Table 7.1. These major enablers and the associated sub-major enablers are classified from the point of perspective of the experts to be interviewed. It was observed from Table 7.1 that the sustainability of the facilities design layout systems was mainly focused on the economy, society, and environment pillars in enablers and sub-enablers. All these enablers either majors or sub-majors including the TBL of sustainability are used to generate the four possible manufacturing systems.

Practically, these relative weights are estimated as from 40% to 50% for cellular manufacturing cells; 30%–40% for functional cells, and 5%–10% for dedicated cells as given in the following Table 7.2 (Garbie and Garbie, 2024a,c).

7.3 ANALYSIS AND ASSESSMENT OF SUSTAINABLE MANUFACTURING SYSTEMS

The given layer of manufacturing is the so-called manufacturing systems (MS), and it is utilized subsequently as a result of the last past layers of manufacturing: core and chosen. The outcome of this layer is represented within the diverse types of manufacturing: functional or process layout "Job Shop" systems, product layout "Flow Job", and focused layout "Cellular" systems (conventional or flexible cells and systems). Although all these manufacturing systems are known as designing manufacturing systems, they are still focusing on the core and chosen layers of manufacturing which were mentioned in previous Chapters 4 and 5. The sustainability level of the MS at any time t, $S_{MS}(t)$ is numerically expressed based on the type of designing facilities layout either Job shop (z_1) or Flow shop (z_2) or Cellular systems (z_3) and at last the assembly lines (either manual or mechanized) (z_4) (Equation 7.1). Embracing one or two of these types of manufacturing systems is broadly suggested especially during the following period and beneath the umbrella of Industry 4.0 [5].

$$S_{MS}(t) = f(z_1, z_2, z_3, \ldots, z_n) = \begin{Bmatrix} z_1 \\ z_2 \\ z_3 \\ z_4 \end{Bmatrix} = \sum_{i=1}^{4} z_i(t)/4 \tag{7.1}$$

where:

$z_i(t)$ in Equation (7.1) is calculated using Equation (7.2)

$$z_i(t) = \frac{z_i(t)\text{current} - z_i(t)\min}{z_i(t)\max - z_i(t)\min} \tag{7.2}$$

where:

$z_i(t)$ is the normalized value of aspects of the property specified with the manufacturing systems (MS) at any time t,

$z_i(t)$current is the existing value of the property specified in any part of the manufacturing systems at any time t,

TABLE 7.1
Enablers of Manufacturing Systems for Sustainability

Major Enablers of Manufacturing System (MS)	Sustainability Pillars Concentration	Sub-Enablers/Incentives/ Drivers/Motivation	Given Manufacturing System Design Types
Globalization and international regulations	Economy, society, and environment	• Supply chain management • Information communication and technology • Energy price • Markets and business models	• Job shop manufacturing systems (z_1) • Flowline (mass) manufacturing systems (z_2) • Cellular manufacturing systems (z_3) • Assembly lines (z_4)
Emerging and contingency matters	Economy, society, and environment	• Technology • Governmental regulations • Natural resources	
Innovation products	Economy and society	• New products • Product development • Mass customization	
Design and reconfiguration system	Economy	• Designing the system • Reconfiguration of the system	
Competitive manufacturing strategies	Economy and environment	• Optimize Complexity • Lean production • Agile manufacturing • Remanufacturing • Recycling	
Performance evaluation	Economy and society	• Product cost • Manufacturing lead time • Productivity • Human resources appraisal • Resource status • Quality level	
organizational management	Economy and society	• Strategic planning • Organizing work • Organizational structure • Leadership style • Staffing • Managing culture	

TABLE 7.2
Relative Weights between Types of Manufacturing Systems [5]

Manufacturing System	Cellular (Focused) System	Functional System-Job Shop	Dedicated System-Flow Shop
Relative weights, %	40%–50%	30%–40%	5%–10%

$z_i(t)$max and $z_i(t)$min are the maximum and minimum values of the property specified in any component of the manufacturing systems at any time t, respectively,

$z_1(t)$ = normalizing job shop manufacturing system level at any time t.

$$z_1(t) = \frac{E^{z1} - I_{min}^{z1}}{I_{max}^{z1} - I_{min}^{z1}} \quad (7.3)$$

E^{z1} = the existing job shop manufacturing system at any time t,

I_{max}^{z1} = Maximum using job shop manufacturing system at any time t,

I_{min}^{z1} = Minimum using job shop manufacturing system at any time t,

$z_2(t)$ = normalizing flow shop manufacturing system level at any time t.

$$z_2(t) = \frac{E^{z2} - I_{min}^{z2}}{I_{max}^{z2} - I_{min}^{z2}} \quad (7.4)$$

E^{z2} = the existing using flow shop manufacturing system at any time t,

I_{max}^{z2} = maximum using flow shop manufacturing system at any time t,

I_{min}^{z2} = minimum using flow shop manufacturing system at any time t,

$z_3(t)$ = normalizing cellular manufacturing system level at any time t.

$$z_3(t) = \frac{E^{z3} - I_{min}^{z3}}{I_{max}^{z3} - I_{min}^{z3}} \quad (7.5)$$

E^{z3} = the existing cellular manufacturing system at any time t,

I_{max}^{z3} = maximum using cellular manufacturing system at any time t,

I_{min}^{z3} = minimum using cellular manufacturing system at any time t,

$z_4(t)$ = normalizing using automated systems rather than manual systems level at any time t.

$$z_4(t) = \frac{E^{z4} - I_{min}^{z4}}{I_{max}^{z4} - I_{min}^{z4}} \quad (7.6)$$

E^{z4} = the existing using automated systems rather than manual systems level at any time t,

I_{max}^{z4} = maximum using automated systems rather than manual systemslevel at any time t,

I_{min}^{z4} = minimum using automated systems rather than manual systemslevel at any time t.

7.4 NUMERICAL EXAMPLE 7.1

This numerical example demonstrates the proposed method for evaluating the sustainable manufacturing index for manufacturing systems. The evaluation focuses on identifying sustainability enablers concerning manufacturing system design and performance measurements, which include existing values and target values (minimum and maximum). Relevant information about these sustainability enablers is presented in Table 7.3 for manufacturing systems.

Concerning the sustainability of manufacturing systems, the same strategy will be taken after as clarified in machine design sustainability and manufacturing

TABLE 7.3
Data for Manufacturing Systems

Indicator (*z*)	Units Performance Measures	Performance Metrics		
			Target Value	
		Existing Value	Max	Min
z_1	Percentage (%) of using job shop manufacturing system in a plant	36	40	30
z_2	Percentage (%) of using flow shop manufacturing system in a plant	8	10	6
z_3	Percentage (%) of using cellular manufacturing system in a plant	46	50	40
z_4	Percentage (%) of automated systems rather than manual systems	75	80	60

TABLE 7.4
Sustainability Index in Manufacturing Systems

Indicator (*z*)	z_1	z_2	z_3	z_4
Normalized index	0.60	0.50	0.60	0.75
$S_{MS}(t)$	0.6125			

processes for normalized indicators and assessed the sustainability indexes by using Equation 7.1 as in previous chapters, as appeared in Table 7.4. For example, the indicator for job shop (z_1), which is evaluated as the percentage (%) of using the job shop manufacturing system at any time by utilizing two values for the target (greatest as 40% and least as 30%) and one existing value as 36%. At that point, the normalized value of using job shop is assessed as 0.60 as given in the following equation and appears in Table 7.3.

$$z_1(t) = \frac{36-30}{40-30} = 0.60$$

The sustainability index of manufacturing systems at any time t, $s_{MS}(t)$, is evaluated by utilizing Equation 7.1, and the results are outlined in Table 7.4 as 0.6125 (61.25%). The sustainability index of manufacturing systems is calculated to be 61.25%. While this value is generally considered moderate, efforts should be made to improve it across all types of manufacturing systems to enhance sustainability.

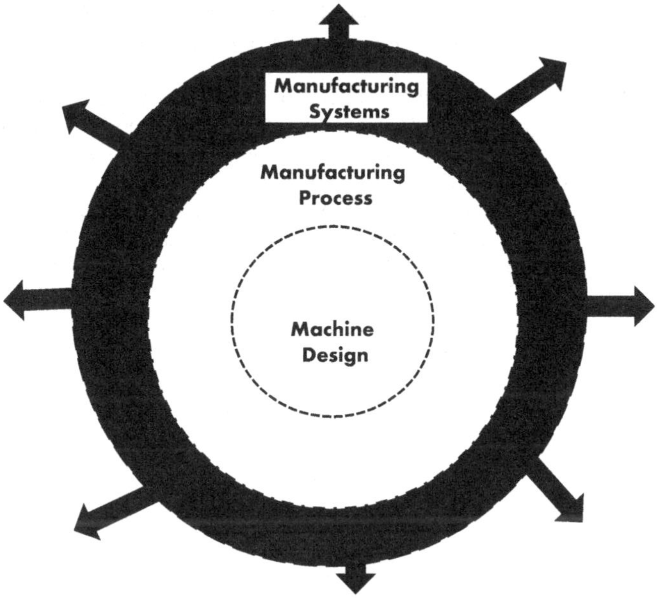

FIGURE 7.3 Outcomes of a given layer (manufacturing systems).

7.5 CONCLUDING MARKS

The third layer of sustainable manufacturing, viewed from an architectural perspective, was proposed and analyzed through the given (final) layer, identified as sustainability in manufacturing systems. The primary objective of this layer was to adopt and evaluate the four major manufacturing system types, encompassing both manufacturing/production and assembly systems.

In this context, common practical adoptions across these manufacturing systems were identified, and the sustainability index of the systems was assessed. The architectural framework was fully realized, culminating in a comprehensive investigation of sustainability within manufacturing systems as the final layer. This integration of sustainability principles in the given layer is illustrated in Figure 7.3.

REFERENCES

Garbie, I., (2013), "DFSME: Design for Sustainable Manufacturing Enterprises (An Economic Viewpoint)", *International Journal of Production Research*, Vol. 51, No. 2, pp. 479–503.

Garbie, I., (2014), "An Analytical Technique to Model and Assess Sustainable Development Index in Manufacturing Enterprises", *International Journal of Production Research*, Vol. 52, No. 16, pp. 4876–4915.

Garbie, I., and Garbie, A., (2022), "A Holistic Perspective of Sustainable Manufacturing: A Novel Conceptual Approach", *International Journal of Industrial and Systems Engineering*, Vol. 41, No. 2, pp. 135–167.

Garbie, I., and Parsaei, H., (2022), *Reconfigurable Manufacturing Enterprises for Industry 4.0*, 1st Edition. CRC Press-Taylor and Francis, Boca Raton, FL, USA.

Garbie, I., and Garbie, A., (2024a) "Toward Smart Manufacturing Systems Incorporating Reconfiguration Issues", *International Journal of Industrial and Systems Engineering*, Vol. 46, No. 1, pp. 1–33.

Garbie, I., and Garbie, M., (2024b), "Sustainability of Manufacturing Enterprises: As Psychological Perspective of Linking Manufacturing Layers", *Proceedings of the Industrial and Systems Engineering Research Conference (IISE Annual Conference and Expo 2024)*, May 18–21, 2024, Montreal, Canada, 6 pp.

Garbie, I., and Garbie, M., (2024c), "Designing Manufacturing Enterprises System for Industry 4.0", *Proceedings of the Industrial and Systems Engineering Research Conference (IISE Annual Conference and Expo 2024)*, May 18–21, 2024, Montreal, Canada, 6 pp.

8 Manufacturing Sustainability

8.1 INTRODUCTION TO MANUFACTURING SUSTAINABILITY

It can be shown in the following Figure 8.1 that there are three different layers of the architectural perspective to sustainable manufacturing (Garbie and Garbie, 2024b). The primary layer is so-called the core and is a dependable machine/product design components. The second layer is the chosen layer, represented by the manufacturing processes either forming or machining with which feasible and convenient manufacturing processes will be adopted. The final layer could be a given one and is utilized to demonstrate which manufacturing system type can be given and suggested (Garbie and Garbie, 2024c). This chapter aims to discover the sustainability index for these layers individually first and the architecture interface between them. In this chapter also, the manufacturing technologies of these three layers will be distinguished and analyzed, and the sustainability index linking these layers through an architecture

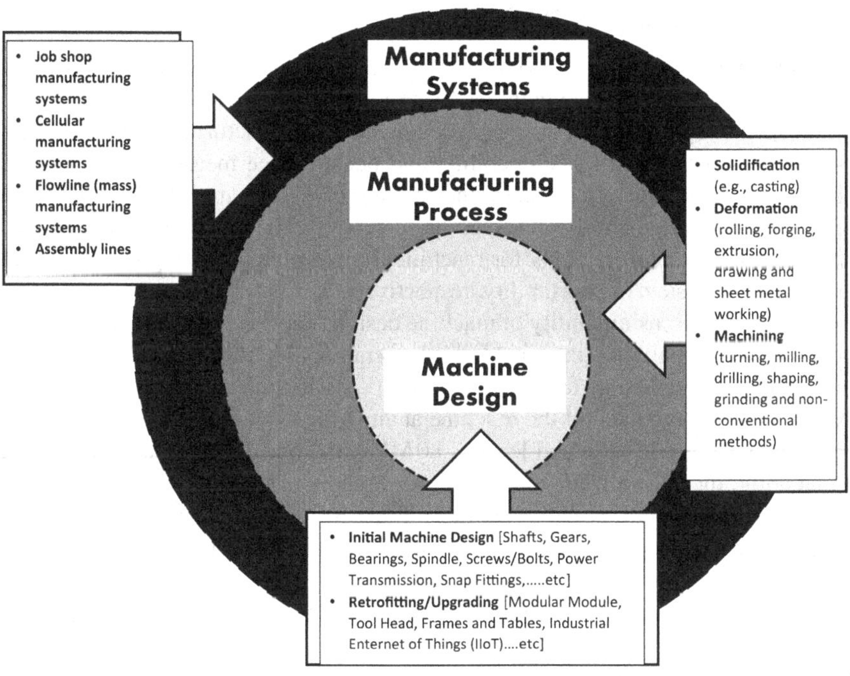

FIGURE 8.1 Layers of sustainable manufacturing in terms of the architecture perspective.

DOI: 10.1201/9781003644835-11

perspective will be proposed and outlined independently and universally. It is useful for implementing Industry 4.0/smart manufacturing (Garbie and Garbie, 2024a).

8.2 ASSESSMENT OF MANUFACTURING SUSTAINABILITY

The index of manufacturing sustainability is based on the sustainability of the three layers of manufacturing: machine/product components design, $S_{\text{MD}}(t)$, manufacturing processes, $S_{\text{MP}}(t)$, and manufacturing systems, $S_{\text{MS}}(t)$.

At that point, the sustainable manufacturing index is scientifically communicated at any time t, $S_M(t)$, as a function of the following Equation (8.1) (Garbie and Garbie, 2024a).

$$S_M(t) = \left[S_{\text{MD}}(t)\right]^{\frac{1}{\sum_{j=1}^{2} w_j(t) S_j(t)}} = \left[S_{\text{MD}}(t)\right]^{\frac{1}{\left[w_{\text{MP}}(t) S_{\text{MP}}(t) + w_{\text{MS}}(t) S_{\text{MS}}(t)\right]}} \quad (8.1)$$

Equation (8.1) is based on the core layer of manufacturing "sustainability of machine/product components design ($S_{\text{MD}}(t)$)" which is expressed as a constant value and later sustainability in manufacturing processes $S_{\text{MP}}(t)$, and manufacturing systems ($S_{\text{MS}}(t)$) as a variable. By this way, Equation (8.1) is considered as an exponential function affecting by the three layers of manufacturing. Therefore, the sustainable manufacturing function at any time t, $S_M(t)$, has the form which includes $S_{\text{MD}}(t) > 0$ mathematically and practically.

8.3 NUMERICAL EXAMPLE 8.1

This numerical illustration is utilized to demonstrate the proposed method or system for evaluating the sustainable manufacturing index on the full manufacturing based on machine design, manufacturing processes, and manufacturing systems through distinguishing enablers of sustainability, and performance measurements (existing values and target values (minimum and maximum)). Relevant information regarding sustainability enablers is presented in Tables 8.1–8.3, which detail the enablers and performance measurements for machine design, manufacturing processes, and manufacturing system supportability, respectively.

Concerning the sustainability of machine design (Table 8.1), the foremost imperative issue is normalizing the enablers of sustainability through their indicators. For example, the indicator for tensile quality (x_1), which is evaluated as the tensile strength in any component of the machine at any time by utilizing two values for the target (greatest as 100 MPa and least as 80 MPa) and one existing value as 90 MPa. At that point, the normalized value of tensile strength is assessed as 0.50 as in given in the following equation and appears in Table 8.4. The total assessed indicators result in machine design (MD) sustainability, which appears in Table 8.4.

$$x_1(t) = \frac{90 - 80}{100 - 80} = 0.50$$

The sustainability index of machine components designs at any time t, $s_{\text{MD}}(t)$, is evaluated by utilizing Equation (5.1), and the results are outlined in Table 8.4 as

TABLE 8.1
Data for Machine/Product Components Design

Indicator (x)	Units Performance Measures	Performance Metrics Existing Value	Target Value Max	Min
x_1	Value of tensile strength in any component at time t, MPa	90	100	80
x_2	Value of compression in any component at time t, MPa	100	105	95
x_3	Value of impact in any component at time t, MPa	70	80	50
x_4	Value of hardness in any component at time t, MPa	195	200	190
x_5	Value of toughness in any component at time t, MPa	145	150	140
x_6	Value of torsion in any rotating component at time t, MPa	143	145	140
x_7	Value of shear stress in any component at time t, MPa	148	150	140
x_8	Value of corrosion in any component at time t, mm/year	0.004	0.01	0.005
x_9	Value of fatigue in any component at time t, MPa	200	250	175
x_{10}	Value of fracture in any component at time t, MPa	450	500	400
x_{11}	Value of creep in any component at time t, mm/year	0.005	0.01	0.004
x_{12}	Value of stiffness in any component at time t, MPa	90	100	85
x_{13}	Value of bending and flexure in any component at time t, MPa	185	200	180
x_{14}	Number of modular modules at time t	6	10	4
x_{15}	Possibility to change the head of a machine at any time t	80	100	70
x_{16}	Number of frames, moving and rotating tables in a machine at any time t	3	4	2
x_{17}	Availability of using the industrial Internet of thing (IIOT) at any time t	80	100	70
x_{18}	Number of sensors used in the machine at any time t	10	15	9
x_{19}	Integration of cyber-physical systems (CPS) at any time t	80	90	70
x_{20}	Availability of using embedded computer systems (ECS) at any time t	70	100	60
x_{21}	Availability of using data analytics (DA) at any time t	90	100	80

0.4171 (41.71%). The sustainable index of machine design breaks even with 41.45%, and this value appears generally little, it must be moved forward in all issues/aspects of machine design, either within the introductory design components or reconfigurable issues or microsystem technology. Concerning the sustainability of manufacturing processes and systems, the same strategy will be taken after as clarified in machine design sustainability for normalized indicators and assessed the sustainability indexes by using Equations (6.1) and (7.1) in previous chapters, as appeared in Tables 8.5 and 8.6.

The index of sustainable manufacturing is evaluated as specified in Equation (8.1). This assessment is based on the sustainability of machine component design, manufacturing processes, and manufacturing systems, with equal weights assigned to the manufacturing processes and manufacturing systems.

TABLE 8.2
Data for Manufacturing Processes

Indicator (y)	Units Performance Measures	Performance Metrics: Existing Value	Target Value: Max	Target Value: Min
y_1	The temperature of the oven at any time t, C	1500	2000	1000
y_2	Usage of energy in casting/solidification at any time t, kW	70	100	50
y_3	Utilization (usage) of materials at any time t, %	97	100	95
y_4	Number of products manufactured during the life of the die at any time, t	120 K	150 K	100 K
y_5	Usage of energy in bulk deformation, kW	200	250	180
y_6	Availability of designing a new die at any time t for bulk deformation	90	95	80
y_7	Usage of energy in sheet metal defamation at any time t, kW	18	20	15
y_8	Availability of designing a new die at any time t for sheet metal working	90	95	80
y_9	Energy consumption used to machine or cut a workpiece at any time t, kW	20	25	10
y_{10}	Availability of configuration and cutting angles of a cutting tool at any time t	95	100	90
y_{11}	Cutting/machining time before replacing the cutting tool at any time t	15	25	12
y_{12}	Availability of cooling (coolant) type at any time t	90	95	85

TABLE 8.3
Data for Manufacturing Systems

Indicator (z)	Units Performance Measures	Performance metrics: Existing Value	Target Value: Max	Target Value: Min
z_1	Percentage (%) of using functional (process) manufacturing system in a plant	35	40	30
z_2	Percentage (%) of using flow manufacturing system in a plant	7	10	5
z_3	Percentage (%) of using cellular manufacturing system in a plant	45	50	40
z_4	Percentage (%) of automated systems rather than manual systems	70	80	50

TABLE 8.4
Sustainability Index in Machine/Product Components Design

Indicator (x)	x_1	x_2	x_3	x_4	x_5	x_6	x_7	x_8	x_9	x_{10}	x_{11}	x_{12}	x_{13}	x_{14}	x_{15}	x_{16}	x_{17}	x_{18}	x_{19}	x_{20}	x_{21}
Normalized index	0.5	0.5	0.67	0.5	0.5	0.6	0.8	0.2	0.33	0.5	0.17	0.33	0.25	0.33	0.33	0.5	0.33	0.17	0.5	0.25	0.5
$S_{MD}(t)$	0.4171																				

TABLE 8.5
Sustainability Index in Manufacturing Processes

Indicator (y)	y_1	y_2	y_3	y_4	y_5	y_6	y_7	y_8	y_9	y_{10}	y_{11}	y_{12}
Normalized index	0.50	0.40	0.40	0.40	0.40	0.67	0.60	0.60	0.60	0.50	0.25	0.50
$S_{MP}(t)$	0.485											

TABLE 8.6
Sustainability Index in Manufacturing Systems

Indicator (z)	z_1	z_2	z_3	z_4
Normalized index	0.50	0.40	0.50	0.67
$S_{MS}(t)$	0.5175			

TABLE 8.7
Effects of the Values of Individual Sustainability on the Whole Manufacturing Sustainability

						$S_M(t)$	
Iteration	$S_{MD}(t)$	$w_{MP}(t)$	$S_{MP}(t)$	$w_{MS}(t)$	$S_{MS}(t)$	**Index**	**100%**
1	0.4145	0.50	0.4850	0.50	0.5175	0.1725	17.25
2	0.5145	0.50	0.4850	0.50	0.5175	0.2655	26.55
3	0.6145	0.50	0.4850	0.50	0.5175	0.3785	37.88
4	0.7145	0.50	0.4850	0.50	0.5175	0.5113	51.13
5	0.8145	0.50	0.4850	0.50	0.5175	0.6640	66.40
6	0.9145	0.50	0.4850	0.50	0.5175	0.8366	83.66
7	0.9545	0.50	0.4850	0.50	0.5175	0.9112	91.12

TABLE 8.8
Effects of the Values of Individual Sustainability on Manufacturing Sustainability

						$S_M(t)$	
Iteration	$S_{MD}(t)$	$w_{MP}(t)$	$S_{MP}(t)$	$w_{MS}(t)$	$S_{MS}(t)$	**Index**	**100%**
1	0.9545	0.50	0.90	0.50	0.90	0.95	95.00

$$S_M(t) = [0.4145]^{\frac{1}{[0.50x0.485+0.50x0.5175]}} = 0.1725$$

This implies that sustainable manufacturing achieves a level of 17.25% when equal relative weights are assigned to the sustainability of manufacturing processes and manufacturing systems. It also highlights that significant effort (both in terms of time and cost) is required to improve the overall sustainability level by 82.75% (100%–17.25%).

Adjusting the individual sustainability index for "machine component design" in each stream will affect the overall sustainable manufacturing index while maintaining equal relative weights and the same values for the sustainability of manufacturing processes and manufacturing systems. These changes are demonstrated in Table 8.7.

As observed from the results in Table 8.7, changes in the index of each sustainability layer, particularly in machine design, result in corresponding changes to the overall sustainable manufacturing index, demonstrating its reliance on these individual streams. Table 8.8 presents the proposed highest values for the sustainability index within each of the three layers.

8.4 CONCLUDING REMARKS

In this chapter, the architecture perspective was proposed and analyzed to connect the three primary layers of sustainable manufacturing. The concepts of the core, chosen, and given layers were introduced as novel and inventive approaches within the manufacturing domain. For each layer, the essential components and aspects were identified, and the sustainability index was assessed independently. Subsequently, a global sustainable manufacturing index was evaluated using a numerical scientific model to demonstrate the interdependence of the manufacturing layers. The analysis revealed that an increase in the sustainability of manufacturing processes or manufacturing systems results in a corresponding increase in global sustainable manufacturing, even when the sustainability of machine design remains constant, and vice versa.

Logically, the architecture perspective interface exists between the layers of manufacturing, even within the structured and technical domain of Industry 4.0. The architecture perspective system plays a pivotal role in sustainable manufacturing, influencing not only the behavior and decision-making of individuals within the manufacturing enterprise but also the dynamics between the layers themselves. To further explore the impact of the architecture perspective system, future studies are recommended to investigate real-world case studies focusing on aspects such as employee motivation, decision-making, innovation and creativity, behavior change, and organizational culture.

REFERENCES

Garbie, I., and Garbie, A., (2024a), "Toward Smart Manufacturing Systems incorporating Reconfiguration Issues", *International Journal of Industrial and Systems Engineering*, Vol. 46, No. 1, pp. 1–33.

Garbie, I., and Garbie, M., (2024b), "Sustainability of Manufacturing Enterprises: As Psychological Perspective of Linking Manufacturing Layers", *Proceedings of the Industrial and Systems Engineering Research Conference (IISE Annual Conference and Expo 2024)*, May 18–21, 2024, Montreal, Canada, 6 pp.

Garbie, I., and Garbie, M., (2024c), "Designing Manufacturing Enterprises System for Industry 4.0", *Proceedings of the Industrial and Systems Engineering Research Conference (IISE Annual Conference and Expo 2024)*, May 18–21, 2024, Montreal, Canada, 6 pp.

Part 4

Sustainable Industry 4.0

9 Sustainable Emerging Technologies

9.1 INTRODUCTION TO EMERGING TECHNOLOGIES

Industry 4.0 has propelled the manufacturing sector toward adopting new concepts centered on digitization, automation, and information and communication technology (ICT). The transition of manufacturing enterprises and systems to Industry 4.0 and its implementation is essential (Garbie and Garbie, 2022). Industry 4.0 and smart manufacturing technologies can be categorized into three main groups: basic, intermediate, and advanced technologies (Zaidi et al., 2022).

1. *Basic technologies* include the Internet of Things (IoT), network infrastructure, wireless devices, digital platforms (e-commerce, social media, messaging platforms, email), and smart devices.
2. *Intermediate technologies* encompass big data, cloud systems, and cybersecurity.
3. *Advanced technologies* feature artificial intelligence (AI), cyber-physical systems (CPS), augmented reality (AR), robotics, and sensors.

The digitalization (DI) of manufacturing supply chains significantly impacts supply chain operations, leading to enhanced productivity and growth (Adeyemia et al., 2024).

In general, emerging technologies are divided into two major categories based on the physical world and the digital world: advanced manufacturing technologies and advanced digital technologies. Emerging advanced digital and manufacturing technologies for Industry 4.0(I4.0) are considered the basic and core layer toward implementing I4.0, but selecting which one is feasible and more convenient for implementation is highly appreciated. Customizing these advanced emerging technologies is different and oriented toward the industry, especially manufacturing industries need more attention from academicians and practitioners based on the nature of the industry itself and its manufacturing processes.

Advanced manufacturing and digital technologies that enable Industry 4.0 include IoT, cyber-physical production systems (CPPS), big data analytics (BDA), DI, cloud computing (CC), cloud manufacturing (CM), cybersecurity (CS), virtual and augmented reality (VR/AR), additive manufacturing (AM), and more (Figure 9.1).

Figure 9.1 illustrates that emerging technologies for Industry 4.0 are divided into three main areas:

DOI: 10.1201/9781003644835-13

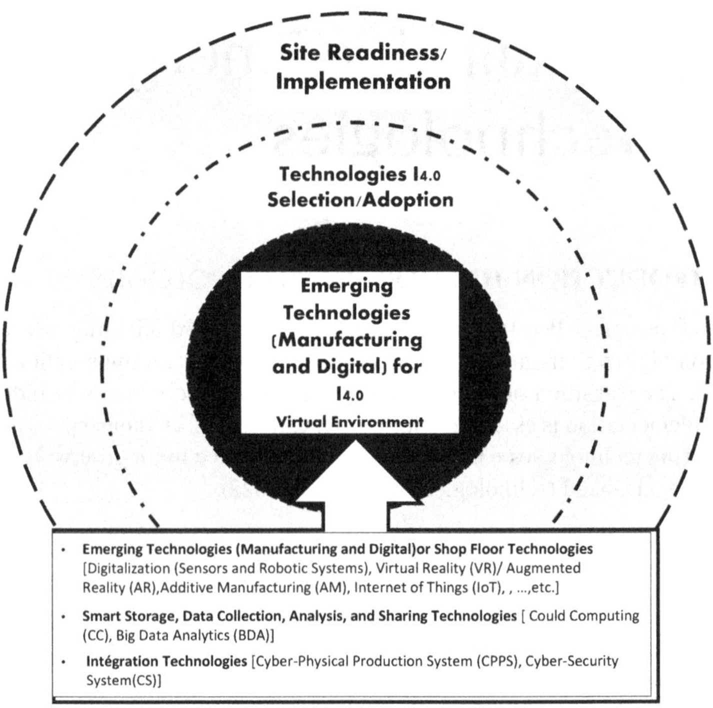

FIGURE 9.1 Emerging technologies (manufacturing and digital) for I4.0 concerning other layers.

1. **Shop Floor Technologies**: DI, VR/AR, and additive manufacturing (AM).
2. **Smart Storage and Data Management**: CC, CM, and BDA.
3. **Integration Technologies**: CPPS and CS

Some advanced manufacturing technologies are classified based on their relevance to the virtual and/or physical environments (Garbie & Parsaei, 2022). These technologies bridge the gap between the real and virtual worlds, functioning as enabling tools for implementing Industry 4.0.

Adopting these cutting-edge technologies has become essential for manufacturing industries to remain competitive. The subsequent sections of this chapter will examine each emerging technology individually.

9.2 COMPONENTS OF EMERGING TECHNOLOGIES

Several developed nations, including Germany, the UK, France, Italy, and the USA, have prioritized funding academic institutions and manufacturing enterprises to initiate research related to implementing Industry 4.0 (Garbie and Parsaei, 2022). These efforts emphasize the critical importance of advanced, customized (enabling) manufacturing technologies in successfully deploying Industry 4.0 frameworks. This section analyzes and discusses the advanced enabling technologies required for implementing

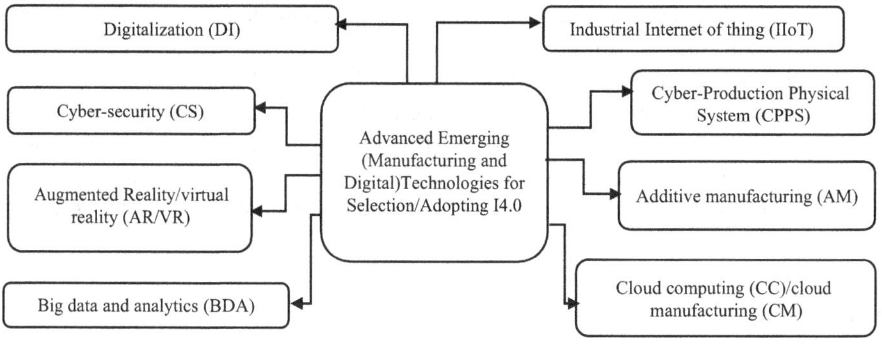

FIGURE 9.2 Emerging technologies (manufacturing and digital) for Industry 4.0.

Industry 4.0, as depicted in Figure 9.2. These technologies are tailored to meet the specific needs of manufacturing industries, providing the foundation for innovation and transformation within the Industry 4.0 paradigm. Emerging technologies span a range of functionalities, from advanced manufacturing techniques to digital tools and integration systems. Each component is integral to the seamless integration of physical and digital processes, enabling enhanced efficiency, sustainability, and competitiveness.

9.2.1 Digitalization or Digital Manufacturing (DI)

DI, also referred to as digital manufacturing (DI), is a process that enables manufacturing technologies such as simulation and virtual reality (VR) to enhance and support production systems. The concept of DI is not a new phenomenon, its roots trace back several decades to the era of Industry 3.0, with the advent of technologies like numerical control (NC) and computerized numerical control (CNC).

Key enabling technologies of DI in manufacturing industries include product description technologies, manufacturing process and control technologies, simulation and virtual reality technologies, and manufacturing data acquisition, storage, and processing technologies. While these technologies have been foundational, they are no longer sufficient for the evolving demands of Industry 4.0. To address the requirements of the next industrial era, additional features and characteristics are urgently needed. These include advanced instruments and devices, sophisticated software, and enhanced engineering management systems, all of which are critical for designing future digital manufacturing processes. The new generation of enabling technologies for DI must incorporate innovations like the IoT and CPPS. These advanced technologies are essential for bridging the gap between existing digital systems and the interconnected, intelligent systems envisioned for Industry 4.0 (Garbie and Parsaei, 2022).

9.2.2 IoT or Industrial Internet of Things (IIoT)

The IoT, or IIoT, is recognized as one of the critical emerging manufacturing technologies within the framework of Industry 4.0. IoT and IIoT enable the interconnection of numerous entities through a global information network. In the context of manufacturing, these interconnected "things" may include materials, sensors, actuators, controllers, robots, human operators, machines, equipment, products, and material handling systems, among others (Garbie and Parsaei, 2022). Through these interconnected networks, the IIoT transforms machines, operators, sensors, devices, raw materials, vehicles, bicycles, and other objects into "smart" components. This enhanced connectivity facilitates seamless communication and integration, fostering more efficient, intelligent, and adaptive manufacturing processes.

9.2.3 Big Data Analytics

BDA is a powerful tool used to process and illustrate data collected from diverse sources. This data is characterized by three primary dimensions: volume, velocity, and variety. Each dimension possesses distinct traits. For instance, volume pertains to the size of the data, velocity relates to the rate at which data is generated and processed, and variety refers to the inclusion of different types of data with varying uses and methods of analysis. One of the main advantages of BDA is its ability to retrieve relevant data across high-volume, high-velocity, and high-variety datasets. Recent advancements in ICTs have significantly increased the generation of big data. Data now originates from numerous sources, including machine controllers, embedded sensors in various devices, CC platforms, mobile devices, and more. However, implementing BDA in manufacturing industries requires substantial infrastructure. Specifically, BDA relies on a robust platform and architecture to facilitate data storage, retrieval, management, processing, and analysis (Garbie et al., 2023; Garbie and Garbie, 2024a,b).

9.2.4 CC or CM

CC or CM is among the most critical industrial applications of Industry 4.0, facilitating the creation of virtual environments for implementing Industry 4.0 technologies such as IIoT, BDA, and DI. CC/CM is defined as a model that enables ubiquitous, convenient, and on-demand network access with minimal management effort or service-provider interaction. This is achieved through internet-incorporated software and platforms accessed via applications and services. CC is a novel technique for processing and managing data, whereby components of the virtual environment in Industry 4.0 are pooled under the umbrella of the internet. For this reason, CC is considered one of the most significant elements of a virtual environment. Its applications extend beyond industrial uses and are commonly employed in public platforms, such as Amazon.com, as well as prominent social media platforms like Facebook and Twitter (Garbie and Parsaei, 2022).

9.2.5 CS and Data Privacy

The IIoT, CPPS, and big data rely on robust CS measures to ensure their protection. Additionally, safeguarding data privacy is a critical obligation. Without securing advanced manufacturing technologies in both virtual and physical environments, the implementation and adoption of Industry 4.0 would face significant challenges. CS is founded on four central pillars: confidentiality, integrity, availability, and data encryption. These pillars collectively ensure that data and systems remain secure, accessible, and trustworthy, thereby supporting the seamless integration of Industry 4.0 technologies (Garbie and Parsaei, 2022).

9.2.6 VR/AR

VR and AR represent a set of technologies that enable users to interact with computers in simulated environments or enhance real-world environments with computer-generated elements. VR immerses the user in a completely simulated digital environment, while AR overlays digital elements onto the real world to augment the user's perception of their surroundings. AR is particularly notable for its ability to collect and integrate information about the physical environment, offering highly flexible applications across various industries. An AR system comprises four key components: visualization technology, a camera, a tracking system, and a user interface. Among these, visual devices are the most recommended technology for AR implementation due to their effectiveness in enhancing the user experience (Garbie and Parsaei, 2022).

9.2.7 Additive Manufacturing (AM)

AM is one of the emerging technologies that has gained prominence over the past decade. As a critical advanced manufacturing technology, AM is essential for implementing Industry 4.0 and serves as a foundational element for creating the physical environment necessary for smart manufacturing. AM is defined as the process of joining materials layer by layer to fabricate objects based on a 3D model. Often referred to as 3D printing, AM is widely used in applications such as producing prototypes and mockups, replacement parts, dental crowns, artificial limbs, and even bridges. Given its role as a physical component in the smart manufacturing environment, AM is highly recommended for next-generation manufacturing. It enables mass customization to produce personalized and region-specific items, supporting the creation of intricate products with advanced attributes, exceptional accuracy, and specialized characteristics (Garbie and Parsaei, 2022).

9.2.8 Cyber-Physical Production Systems

The adoption of numerical control (NC) and computerized numerical control (CNC) technologies are increasingly urgent in the evolving manufacturing industry environment. However, due to the high cost of CNC machines, a practical alternative is to transform existing machinery into CPPS through the retrofitting process. This approach enables older machines to integrate seamlessly into the smart manufacturing ecosystem. Standardizing the retrofitting process is critical as the industrial sector transitions into the Fourth Industrial Revolution (Industry 4.0). This revolution is characterized by machines functioning as CPPS, connected to IoT networks, facilitating enhanced automation, connectivity, and efficiency

9.3 ANALYSIS AND ASSESSMENT OF SUSTAINABLE EMERGING TECHNOLOGIES (ET)

As illustrated in Figures 9.1 and 9.2, the Emerging Technologies (ET) of Industry 4.0 and Smart Manufacturing are divided into three fundamental sub-streams: Emerging Manufacturing Technologies (or Shop Floor Technologies), Smart Storage, Data Collection, Analysis, and Sharing Technologies, and Integration Technologies. Each sub-stream is characterized by distinct technological features and capabilities. Together, these sub-streams comprehensively represent the advanced technologies that underpin Industry 4.0, highlighting their critical role in shaping the future of smart manufacturing. The I4.0 technologies for the ET will include DI (e.g., information and communication technology, Sensors, and Robotic Systems) (f_1), Industrial Internet of Things (f_2), Big Data Analytics (f_3), Cloud Computing/Cloud Manufacturing (f_4), Cyber-Security System (f_5), Virtual Reality/ Augmented Reality (f_6), Additive Manufacturing (f_7), Cyber-Production Physical System (f_8),…f_n, etc. The sustainability level of the Industry 4.0/smart manufacturing concerns at any time t for the core stream "emerging technologies for I4.0", $S_{\text{ET}}(t)$, can be numerically communicated as a work of all past technologies as appeared within the following Equation (9.1).

$$S_{\text{ET}}(t) = f(f_1, f_2, f_3, \ldots f_n) = \left\{ \begin{array}{c} f_1 \\ f_2 \\ f_3 \\ \text{—} \\ f_n \end{array} \right\} = \sum_{i=1}^{n} f_i(t)/n \tag{9.1}$$

where:

$f_i(t)$ in Equation (9.1) can be calculated using the following Equation (9.2)

$$f_i(t) = \frac{f_i(t)\text{current} - f_i(t)\text{min}}{f_i(t)\text{max} - f_i(t)\text{min}} \tag{9.2}$$

where:

$f_i(t)$ is the normalized value of technology specified with the emerging technology (ET) at any time t,

$f_i(t)$current is the existing value of technology specified with the emerging technology (ET) at any time t,

$f_i(t)$max and $f_i(t)$min are the maximum and minimum values of technology specified with the emerging technology (ET) at any time t, respectively,

n is the number of technologies in the emerging technology (ET),

$f_1(t)$ = normalizing value of DI level at a given time t.

$$f_1(t) = \frac{E^{f_1} - I^{f_1}_{\text{min}}}{I^{f_1}_{\text{max}} - I^{f_1}_{\text{min}}} \tag{9.3}$$

E^{f_1} = the existing value of DI level at a given time t,

$I_{max}^{f_1}$ = maximum value of DI level at a given time t,
$I_{min}^{f_1}$ = minimum value of DI level at a given time t,
$f_2(t)$ = normalizing value of the IIoT level at a given time t.

$$f_2(t) = \frac{E^{f_2} - I_{min}^{f_2}}{I_{max}^{f_2} - I_{min}^{f_2}} \tag{9.4}$$

E^{f_2} = the existing value of the (IIoT) level at a given time t,
$I_{max}^{f_2}$ = maximum value of the (IIoT) level at a given time t,
$I_{min}^{f_2}$ = minimum value of the (IIoT) level at a given time t,
$f_3(t)$ = normalizing value of BDA level at any time t.

$$f_3(t) = \frac{E^{f_3} - I_{min}^{f_3}}{I_{max}^{f_3} - I_{min}^{f_3}} \tag{9.5}$$

E^{f_3} = the existing value of (BDA) level at a given time t,
$I_{max}^{f_3}$ = maximum value of (BDA) level at a given time t,
$I_{min}^{f_3}$ = minimum value of (BDA) level at a given time t,
$f_4(t)$ = normalizing value of CC/CM level at a given time t.

$$f_4(t) = \frac{E^{f_4} - I_{min}^{f_4}}{I_{max}^{f_4} - I_{min}^{f_4}} \tag{9.6}$$

E^{f_4} = the existing value of (CC/CM) level at a given time t,
$I_{max}^{f_4}$ = maximum value of (CC/CM) level at a given time t,
$I_{min}^{f_4}$ = minimum value of (CC/CM) level at a given time t,
$f_5(t)$ = normalizing value of cyber security (CS) level at a given time t.

$$f_5(t) = \frac{E^{f_5} - I_{min}^{f_5}}{I_{max}^{f_5} - I_{min}^{f_5}} \tag{9.7}$$

E^{f_5} = the existing value of (CS) level at a given time t,
$I_{max}^{f_5}$ = maximum value of (CS) level at a given time t,
$I_{min}^{f_5}$ = minimum value of (CS) level at a given time t,
$f_6(t)$ = normalizing value of VR/AR level at a given time t.

$$f_6(t) = \frac{E^{f_6} - I_{min}^{f_6}}{I_{max}^{f_6} - I_{min}^{f_6}} \tag{9.8}$$

E^{f_6} = the existing value of (VR/AR) level at a given time t,
$I_{max}^{f_6}$ = maximum value of (VR/AR) level at a given time t,
$I_{min}^{f_6}$ = minimum value of (VR/AR) level at a given time t,
$f_7(t)$ = normalizing value of (AM) level at a given time t.

$$f_7(t) = \frac{E^{f_7} - I_{min}^{f_7}}{I_{max}^{f_7} - I_{min}^{f_7}} \tag{9.9}$$

E^{f_7} = the existing value of (AM) level at a given time t,
$I^{f_7}_{max}$ = maximum value of (AM) level at a given time t,
$I^{f_7}_{min}$ = Minimum value of (AM) level at a given time t,
$f_8(t)$ = normalizing value of (CPPS) level at a given time t.

$$f_8(t) = \frac{E^{f_8} - I^{f_8}_{min}}{I^{f_8}_{max} - I^{f_8}_{min}} \tag{9.10}$$

E^{f_8} = the existing value of (CPPS) level at a given time t,
$I^{f_8}_{max}$ = maximum value of (CPPS) level at a given time t,
$I^{f_8}_{min}$ = minimum value of (CPPS) level at a given time t.

9.4 NUMERICAL EXAMPLE 9.1

This numerical illustration is used to demonstrate the proposed methodology for assessing the sustainability index of Industry 4.0/Smart Manufacturing from a psychological perspective, based on the available emerging manufacturing technologies. The distinguishing indicators of sustainability issues, along with performance measurements—comprising existing values and target values (minimum and maximum)—are presented in Table 9.1 for emerging manufacturing technologies.

Concerning the sustainability of emerging manufacturing technologies (Table 9.1), the primary imperative is the normalization of these technologies' sustainability through their respective indicators. For example, the indicator for DI ($f1$), is evaluated as the percentage of DI at any time t by utilizing two values for the target (greatest as 100% and least as 80%) and one existing value as 85%. At this stage, the normalized value of DI is evaluated as 0.25, as shown in the following equation and summarized in Table 9.2. The cumulative assessment of indicators determines the sustainability level of emerging manufacturing technologies, with the results also presented in Table 9.2.

$$f_1(t) = \frac{85-80}{100-80} = 0.25$$

The sustainability index of emerging manufacturing technologies (ET) at any time t, $s_{ET}(t)$ is evaluated by utilizing Equation (9.1), and the results are outlined in Table 9.2 as 0.4475 (44.75%). The sustainability index of emerging manufacturing technologies is calculated at 44.75%. While this value is relatively low, it underscores the need for improvement across all indicators of emerging manufacturing technologies.

9.5 CONCLUDING REMARKS

The first stream of sustainable Industry 4.0/Smart Manufacturing, viewed from a psychological perspective was proposed and analyzed through its core stream, identified as an emerging advanced manufacturing technology for Industry 4.0. The concepts

TABLE 9.1
Data for the Emerging Manufacturing Technologies

Indicator (f)	Units Performance Measures (%)	Performance Metrics		
		Existing Value	Target Value	
			Max	Min
f_1	Percentage (%) of Digitalization (DI) at a given time t	85	100	80
f_2	Percentage (%) of (IIoT) at a given time t	95	100	90
f_3	Percentage (%) of (BDA) at a given time t	85	90	80
f_4	Percentage (%) of (CC/CM) at a given time t	80	95	75
f_5	Percentage (%) of (CSS) at a given time t	95	100	90
f_6	Percentage (%) of (VR)/AR) at a given time t	80	90	75
f_7	Percentage (%) of (AM) at a given time t	95	100	80
f_8	Percentage (%) of (CPPS) at a given time t	95	100	90

TABLE 9.2
Sustainability Index in Emerging Manufacturing Technologies

Indicator (f)	f_1	f_2	f_3	f_4	f_5	f_6	f_7	f_8
Normalized Index	0.25	0.50	0.50	0.25	0.50	0.33	0.75	0.50
$S_{ET}(t)$	0.4475							

of this core stream, based on eight major advanced and customized technologies, were examined and presented as a novel and innovative approach within the Industry 4.0/Smart Manufacturing framework. The key enablers of these emerging advanced manufacturing technologies were identified and analyzed, and the sustainability index was assessed. It was observed that the core stream of sustainability in Industry 4.0/Smart Manufacturing serves as the foundation, influencing and initializing sustainability throughout the entire Industry 4.0/Smart Manufacturing system. The psychological system plays a pivotal role in Industry 4.0/Smart Manufacturing by connecting the core stream to subsequent streams, referred to as "chosen" and "given", as depicted in Figure 9.3.

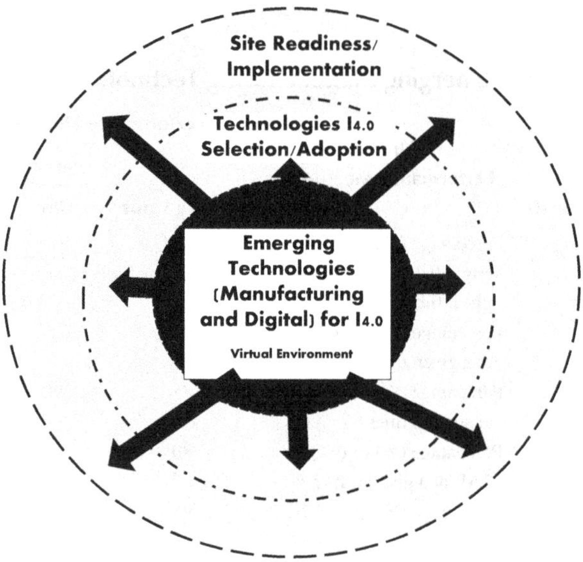

FIGURE 9.3 Affecting core layer (emerging technologies) on the chosen and given layers.

REFERENCES

Adeyemia, O. A., Pintob, P. M. G., Sunmolac, F., Aibinud, A. M., Okesolae, J. O., and Adeyemi, E. O., (2024), "Towards the Adoption of Industry 4.0 Technologies in the Digitalization of Manufacturing Supply Chain", *Procedia Computer Science,* Vol. 232, pp. 337–347.

Garbie, I., and Garbie, A., (2024a), "Toward Smart Manufacturing Systems Incorporating Reconfiguration Issues", *International Journal of Industrial and Systems Engineering*, Vol. 46, No. 1, pp. 1–33.

Garbie, I., and Garbie, M., (2024b), "Sustainability of Manufacturing Enterprises: As Psychological Perspective of Linking Manufacturing Layers", *Proceedings of the Industrial and Systems Engineering Research Conference (IISE Annual Conference and Expo 2024)*, May 18–21, 2024, Montreal, Canada, 6 pp.

Garbie, I, and Parsaei, H., (2022), *Reconfigurable Manufacturing Enterprises for Industry 4.0*, 1st Edition. CRC Press-Taylor and Francis, Boca Raton, FL, USA.

Garbie, I., Parsaei, H., and Garbie, M., (2023), "Site Readiness Index for Implementing Industry 4.0 based on Design Principles and Emerging Manufacturing Technologies", *Proceedings of the Industrial and Systems Engineering Research Conference (IISE Annual Conference and Expo 2023)*, May 20–23, 2023, New Orleans, Louisiana, USA, 6 pp.

Zaidi, N. I. N., Nazmudeen, M. S., and Mohiddin, F., (2022), "Identifying Factors Contributing to the Level of Industry 4.0 Technologies Adoption among SMEs in Different Countries", *International Journal of Engineering and Technology*, Vol. 14, No. 3, pp. 43–51.

10 Sustainable Technologies Selection

10.1 INTRODUCTION TO SELECTION TECHNOLOGIES

The concepts of adopting and implementing Industry 4.0/smart manufacturing are often used interchangeably, but they differ in scope and intent (Zhou and Zheng, 2023). Adoption refers to the decision-making process involved in introducing advanced digital or manufacturing technologies, whereas implementation pertains to their deployment and practical utilization. The adoption of Industry 4.0/smart manufacturing technologies is crucial for enhancing performance management within industrial organizations.

This involves measuring, evaluating, and improving employee performance (Pawar et al., 2023). Emerging technologies significantly influence Human Resource Management (HRM), particularly in the realm of digital HR practices (HR4.0) and performance management systems. For example, Zaidi et al. (2022) investigated the adoption of Industry 4.0 technologies across manufacturing companies in both developing and developed countries, identifying factors that influence adoption levels through qualitative and descriptive methodologies.

At its core, Industry 4.0 is defined by the digital transformation of manufacturing industries. The integration of information and communication technologies (ICT) with emerging industrial technologies is fundamental to this transformation. These technologies drive improvements in organizational performance metrics, such as innovation, operational efficiency, sustainability, and profitability (Zhou and Zheng, 2023). For instance, Ooi et al. (2023) highlighted the role of Industry 4.0 technologies in enhancing organizational performance through lean production practices. Similarly, Tortorella et al. (2024) examined the combined adoption of Industry 4.0 technologies and total productive maintenance (TPM) practices across manufacturing firms in 16 countries.

The emergence of Industry 4.0, driven by advancements in ICT, represents a paradigm shift in manufacturing. As the backbone of industrialized society, manufacturing industries must adapt their structures, processes, and products to meet the demands of intelligent manufacturing (Garbie, 2013, 2014). Garbie and Garbie (2022) emphasized the need for a systematic and comprehensive approach to selecting Industry 4.0 technologies, ensuring alignment with sustainability objectives.

The selection of manufacturing technologies for Industry 4.0 involves mapping design principles (DP) to emerging technologies, forming the "chosen stream" within the broader framework of achieving sustainability in smart manufacturing. This stream focuses on aligning Industry 4.0 principles with emerging technologies to facilitate implementation. There is a strong psychological connection between the chosen stream and the surrounding layers—the "core stream" preceding it and the "given stream" that follows. This connection reflects a site readiness framework for implementing Industry 4.0 technologies (Figure 10.1). The systemic procedure for

DOI: 10.1201/9781003644835-14

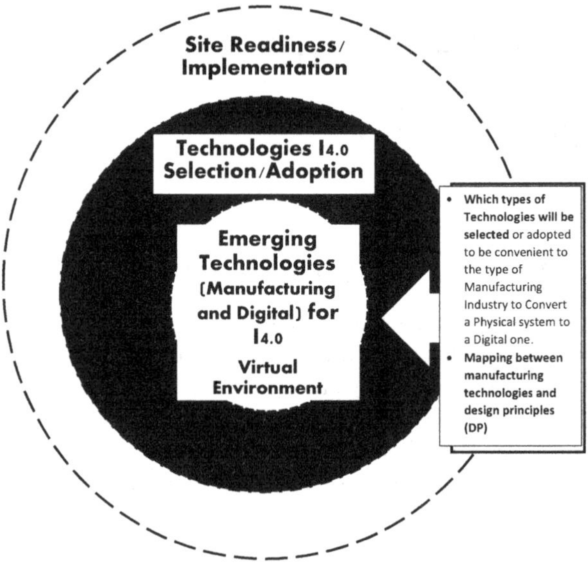

FIGURE 10.1 Technological selection for I4.0 concerning other layers.

technology selection begins by identifying suitable emerging technologies that meet the design principles of Industry 4.0. The chosen stream, positioned between the core and given streams, plays a pivotal role in the transition process, bridging foundational principles with practical readiness for implementation.

10.2 DESIGN PRINCIPLES OF INDUSTRY 4.0

To achieve Industry 4.0 in manufacturing enterprises, six core design principles must be considered and integrated (Gilchrist, 2016; Qin and Cheng, 2017; Garbie and Garbie, 2020a–c). These principles form the foundation for enabling smart manufacturing processes and systems.

10.2.1 Interoperability (IN)

Interoperability is essential for seamless interaction and flexible collaboration among all members of the industrial environment. It is achieved through four levels: operational, systematical, technical, and semantic. These levels ensure that systems, machines, and human operators can effectively communicate and coordinate activities within Industry 4.0 frameworks.

10.2.2 Virtualization (Virtual Reality) (VI)

Virtualization connects the physical world with virtual models through augmented reality technologies. This principle allows the simulation and testing of processes in a virtual environment without affecting real-world systems. Virtualization helps organizations identify potential inefficiencies and make improvements before actual implementation.

10.2.3 Decentralization (DE)

Decentralization enables autonomous decision-making across all members of the supply chain. By dividing enterprises into smaller, manageable components, decentralization enhances the quality and efficiency of operations. It is a critical approach for implementing Industry 4.0, as it empowers individual elements of the system to operate independently while contributing to overall goals (Garbie and Parsaei, 2022).

10.2.4 Real-Time Capability (RT)

Real-time capability is a defining characteristic of Industry 4.0. It involves monitoring manufacturing processes, systems, and supply chain activities while providing immediate feedback from actuators to sensors. This real-time exchange of information enhances responsiveness, reduces downtime, and improves operational efficiency.

10.2.5 Service Orientation (SO)

Service orientation leverages data science and big-data-driven analytics to enable smart systems and services. The Industrial Internet of Things (IIoT), such as the Internet of Services (IoS), facilitates the development of smart factories, products, logistics, grids, and cities. Cyber-Physical Production Systems (CPPSs) play a crucial role in creating these services, providing the infrastructure for adopting Industry 4.0 technologies (Lins and Oliveira, 2020; Garbie and Parsaei, 2022).

10.2.6 Modularity (MO)

Modularity is a key principle for fostering flexibility and agility within manufacturing enterprises. It allows organizations to adapt to changing circumstances by breaking down systems into modular components that can be independently modified or replaced. This principle enhances the adaptability and resilience of smart manufacturing systems (Garbie et al., 2008a,b; Garbie and Parsaei, 2022). Figure 10.2 illustrates the design principles of Industry 4.0.

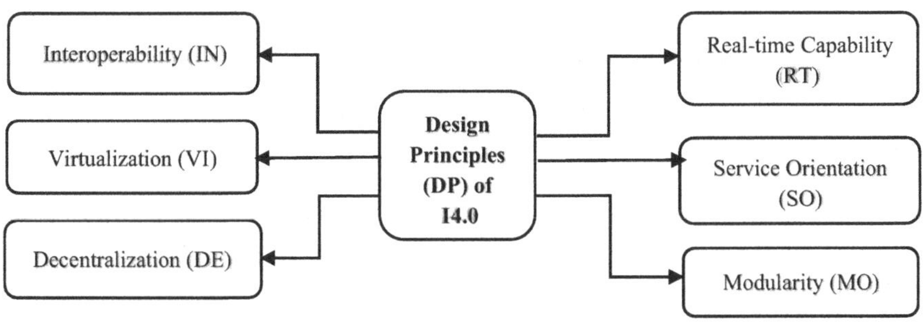

FIGURE 10.2 Design principles (DP) for Industry 4.0.

10.3 MAPPING BETWEEN EMERGING TECHNOLOGIES AND DESIGN PRINCIPLES

Matching emerging technologies for Industry 4.0 with its design principles is essential for understanding and addressing what will be needed to achieve sustainable smart manufacturing. As discussed in Chapter 9, eight or, in some cases, nine advanced manufacturing technologies have been identified within the "core stream". These are supplemented by operational technologies and are linked to the six design principles outlined in Section 10.2.

10.3.1 EMERGING TECHNOLOGIES CATEGORIZATION

Emerging technologies for Industry 4.0 can be categorized into two primary sections:

1. **Digital Technologies**:
   - **Internet of Things (IoT)**: Enables seamless communication and data exchange between devices and systems.
   - **Cyber-Physical Systems (CPSs)**: Integrates physical processes with digital computation and networking.
   - **Cloud Computing/Cloud Manufacturing**: Offers scalable storage and processing power for data-driven decision-making.
   - **Big Data Analytics**: Facilitates insights into performance optimization and predictive maintenance.
   - **Virtual/Augmented Reality**: Supports real-time simulations and enhanced visualization of processes.
2. **Physical Technologies**:
   - **Additive Manufacturing**: Revolutionizes production methods with 3D printing capabilities.
   - **Cybersecurity**: Protects data integrity and systems against cyber threats.
   - **Advanced Manufacturing Solutions**: Includes robotics and automation technologies to enhance efficiency.
   - **Horizontal and Vertical Integration**: Connects systems across organizational levels and value chains for cohesive operations.

10.3.2 MAPPING EMERGING TECHNOLOGIES TO DESIGN PRINCIPLES

The integration of emerging technologies with design principles is vital for aligning technological advancements with organizational goals and sustainability objectives. As illustrated in Figure 10.3, these technologies and principles are interdependent, with digital technologies primarily supporting interoperability, virtualization, and real-time capability, while physical technologies often align with modularity, decentralization, and service orientation. In this section, the mapping between "emerging technologies" and "design principles" will be discussed in detail, as depicted in Figure 10.4. This mapping provides a framework for understanding how specific technologies enable the implementation of Industry 4.0 design principles, ensuring a holistic approach to smart manufacturing.

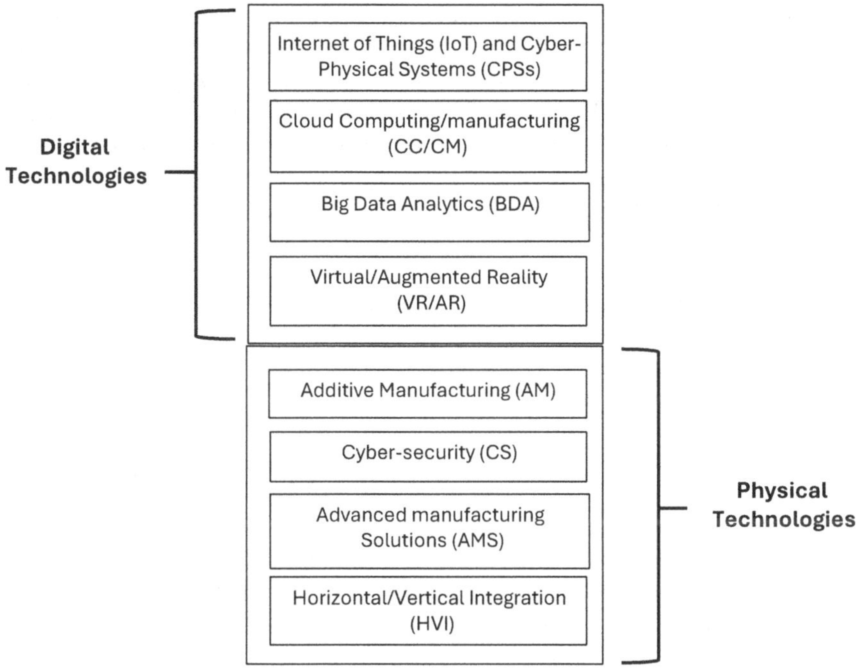

FIGURE 10.3 Digital technologies versus physical technologies in Industry 4.0.

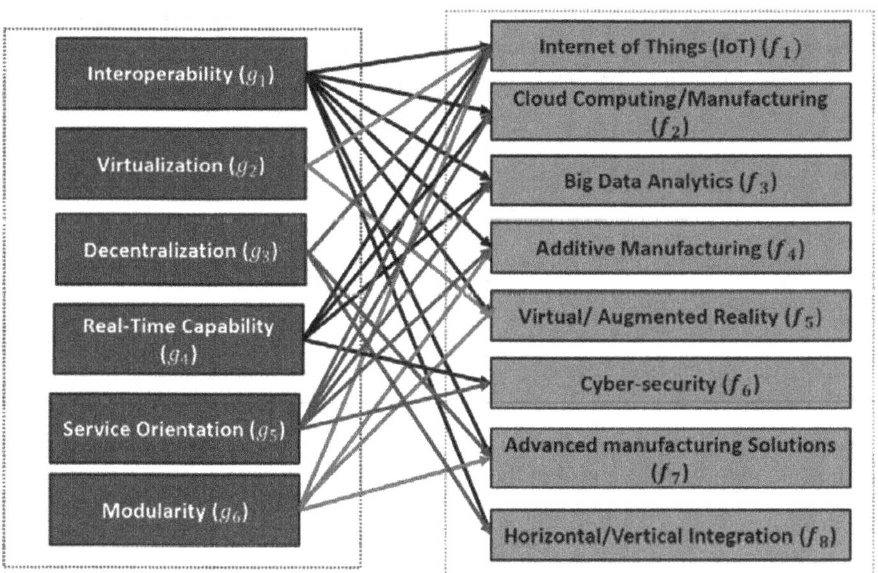

FIGURE 10.4 Illustrates the detailed mapping between emerging technologies and the design principles of Industry 4.0, highlighting their interdependencies and contributions to achieving a sustainable smart manufacturing ecosystem.

In this section, the mapping between "emerging technologies" and "design principles" for Industry 4.0 will be introduced and illustrated, as shown in Figure 10.4. This mapping demonstrates how specific emerging technologies aWlign with and support the realization of Industry 4.0's design principles, ensuring effective implementation of smart manufacturing. As depicted in Figure 10.4, each design principle is matched with the relevant emerging technologies. For example, the design principle of interoperability is enabled by technologies such as:

- Internet of Things (IoT)
- Cloud Computing/Cloud Manufacturing
- Big Data Analytics
- Additive Manufacturing
- Virtual/Augmented Reality
- Advanced Manufacturing Solutions
- Horizontal and Vertical Integration

The alignment of each design principle with its corresponding emerging technologies is determined through insights from academic experts and industrial professionals. This expertise ensures that the matching is grounded in practical application and theoretical understanding, providing a robust framework for technology selection and implementation within Industry 4.0.

It is important to express this mapping by numerically mathematical equation to clarify for readers how the mapping is done as the following Equation (10.1) for interoperability (g_1), for example, as described before as a function of the Internet of Things (f_1), cloud computing/cloud manufacturing (f_2), big data analytics (f_3), additive manufacturing $\left(f_4\right)$, virtual/augmented reality $\left(f_5\right)$, advanced manufacturing solutions $\left(f_7\right)$, and horizontal/vertical integration $\left(f_8\right)$.

$$g_1 = f(f_1, f_2, f_3, f_4, f_5, f_7, f_8) = \left\{ \begin{array}{c} f_1 \\ f_2 \\ f_3 \\ f_4 \\ f_5 \\ f_7 \\ f_8 \end{array} \right\} = \frac{f_1 + f_2 + f_3 + f_4 + f_5 + f_7 + f_8}{7} \quad (10.1)$$

For virtualization (g_2), it is based on the Internet of Things (f_1) and virtual reality/augmented reality (f_5) as given in the following Equation (10.2).

$$g_2 = f(f_1, f_5) = \left\{ \begin{array}{c} f_1 \\ f_5 \end{array} \right\} = \frac{f_1 + f_5}{2} \quad (10.2)$$

For achieving decentralization (g_3), the Internet of Things (f_1), advanced manufacturing solutions (f_7), and horizontal/vertical integration (f_8) are the most important emerging technologies as given in the following Equation (10.3).

$$g_3 = f(f_1, f_8) = \left\{ \begin{array}{c} f_1 \\ f_7 \\ f_8 \end{array} \right\} = \frac{f_1 + f_7 + f_8}{3} \tag{10.3}$$

For real-time capability (g_4), the Internet of Things (f_1), cloud computing/cloud manufacturing (f_2), big data analytics (f_3), and cyber-security (f_6) are the most important emerging technologies for monitoring real-time capability as shown in the following Equation (10.4).

$$g_4 = f(f_1, f_2, f_3, f_6) = \left\{ \begin{array}{c} f_1 \\ f_2 \\ f_3 \\ f_6 \end{array} \right\} = \frac{f_1 + f_2 + f_3 + f_6}{4} \tag{10.4}$$

Concerning service orientation (g_5), the Internet of Things (f_1), cloud computing/cloud manufacturing (f_2), big data analytics (f_3), additive manufacturing (f_4), and cyber security(f_6) are the most important emerging technologies for monitoring service orientation as shown in the following Equation (10.5).

$$g_5 = f(f_1, f_2, f_3, f_4, f_6) = \left\{ \begin{array}{c} f_1 \\ f_2 \\ f_3 \\ f_4 \\ f_6 \end{array} \right\} = \frac{f_1 + f_2 + f_3 + f_4 + f_6}{5} \tag{10.5}$$

For modularity (g_6), it is based on the Internet of Things (f_1), additive manufacturing (f_4), virtual reality/augmented reality (f_5), and advanced manufacturing solutions (f_7), as shown in the following Equation (10.6).

$$g_6 = f(f_1, f_4, f_5, f_7) = \left\{ \begin{array}{c} f_1 \\ f_4 \\ f_5 \\ f_7 \end{array} \right\} = \frac{f_1 + f_4 + f_5 + f_7}{4} \tag{10.6}$$

Concerning $f_1, f_2, f_3, f_4, f_5, f_6, f_7, f_8$, they are estimated in the previous chapter (Chapter 9) by Equations (9.3)–(9.10), respectively

10.4 ANALYSIS AND ASSESSMENT OF SUSTAINABLE TECHNOLOGIES 4.0 SELECTION (TS)

The technologies I4.0 selection (TS) is based on the design principles (DP) given in Section (10.3), and the core layer is mainly dependent on the types of emerging technologies as illustrated in Figure 10.4. The sustainability of TS depends on design principles and their associated emerging technologies. These design principles are interoperability (g_1), virtualization (g_2), decentralization (g_3), real-time capability (g_4), service orientation (g_5), and modularity (g_6). At that point, the sustainability of the Industry 4.0/smart manufacturing concerns technologies selection (TS) at any time t, $S_{\text{TS}}(t)$, is numerically defined as the equation as a function (10.7) of the previous technologies.

$$S_{\text{TS}}(t) = f(g_1, g_2, g_3, \ldots g_6) = \left\{ \begin{array}{c} g_1 \\ g_2 \\ g_3 \\ \text{—} \\ g_6 \end{array} \right\} = \sum_{i=1}^{6} W_i \, g_i(t) \tag{10.7}$$

Equation (10.7) is quantitatively rewritten to evaluate the sustainability index for emerging technologies selection for Industry 4.0 as shown in Equation (10.8)

$$S_{\text{TS}}(t) = W_{g1} g_1(t) + W_{g2} g_2(t) + W_{g3} g_3(t) + W_{g4} g_4(t) + W_{g5} g_5(t) + W_{g6} g_6(t) \tag{10.8}$$

where:

$S_{\text{TS}}(t)$ = sustainability index of emerging technologies selection for Industry 4.0 based on mapping between emerging technologies and design principles at any time t,

$g_1(t)$ = percentage (%) of interoperability at a given time t,

$g_2(t)$ = percentage (%) of virtualization at a given time t,

$g_3(t)$ = percentage (%) of decentralization at a given time t,

$g_4(t)$ = percentage (%) of using real-time capability at a given time t,

$g_5(t)$ = percentage (%) of using service orientation at a given time t,

$g_6(t)$ = percentage (%) of modularity at a given time t,

The symbols W_{g1}, W_{g2}, W_{g3}, W_{g4}, W_{g5} and W_{g6} are the relative weights of interoperability, virtualization, decentralization, real-time capability, service orientation, and modularity, respectively.

10.5 NUMERICAL EXAMPLE 10.1

This numerical illustration demonstrates the proposed methodology for estimating the sustainability index of Industry 4.0 technological selection, matching the design principles (DP) from an architecture perspective. The approach is based on the selection of emerging manufacturing technologies specific to Industry 4.0. The methodology considers distinguishing indicators of sustainability issues and performance measurements. These indicators include existing values and target values (minimum

TABLE 10.1
Customizing Manufacturing Technologies

		Performance Metrics		
	Units		Target Value	
Indicator (f)	Performance Measures (%)	Existing Value	Max	Min
f_1	Percentage (%) of the (IIoT) at a given time t	90	95	85
f_2	Percentage (%) of the (CC/CM) at a given time t	90	100	85
f_3	Percentage (%) of the (BDA) at a given time t	85	95	80
f_4	Percentage (%) of the (AM) at a given time t	85	95	75
f_5	Percentage (%) of the (VR/AR) at a given time t	80	90	70
f_6	Percentage (%) of the (CSS) at a given time t	90	100	80
f_7	Percentage (%) of advanced manufacturing solutions at a given time t	90	100	85
f_8	Percentage (%) of horizontal/vertical integration at a given time t	90	95	85

TABLE 10.2
Normalizing Values in Emerging Manufacturing Technologies

Indicator (f)	f_1	f_2	f_3	f_4	f_5	f_6	f_7	f_8
Normalized index	0.50	0.33	0.50	0.50	0.50	0.50	0.33	0.50

and maximum), which are outlined in Table 10.1. This table serves as a customizable framework for aligning manufacturing technologies selection sustainability objectives with the design principles (DP). The total assessed indicators result in emerging manufacturing technologies sustainability, which appears in Table 10.2 for f_1, f_2, f_3, f_4, f_5, f_6, f_7, and f_8.

For mapping the design principles and emerging technologies for Industry 4.0, Equations (10.1)–(10.6), for interoperability, virtualization, decentralization, real-time capability, service orientation, and modularity, respectively as follows:

$$g_1 = \frac{f_1 + f_2 + f_3 + f_4 + f_5 + f_7 + f_8}{7} = \frac{0.50 + 0.33 + 0.50 + 0.50 + 0.50 + 0.33 + 0.50}{7}$$
$$= 0.4514$$

$$g_2 = \frac{f_1 + f_5}{2} = \frac{0.50 + 0.50}{2} = 0.50$$

$$g_3 = \frac{f_1 + f_7 + f_8}{3} = \frac{0.50 + 0.33 + 0.50}{3} = 0.4433$$

$$g_4 = \frac{f_1 + f_2 + f_3 + f_6}{4} = \frac{0.50 + 0.33 + 0.50 + 0.50}{4} = 0.4575$$

$$g_5 = \frac{f_1 + f_2 + f_3 + f_4 + f_6}{5} = \frac{0.50 + 0.33 + 0.50 + 0.50 + 0.50}{5} = 0.4660$$

$$g_6 = \frac{f_1 + f_4 + f_5 + f_7}{4} = \frac{0.50 + 0.50 + 0.50 + 0.33}{4} = 0.4575$$

Equation (10.8) is used to assess the sustainability index for emerging technologies selection for Industry 4.0 as follows after estimating the relative weights between interoperability, virtualization, decentralization, real-time capability, service orientation, and modularity by using the Analytical Hierarchy Processes as shown in Figure 10.5.

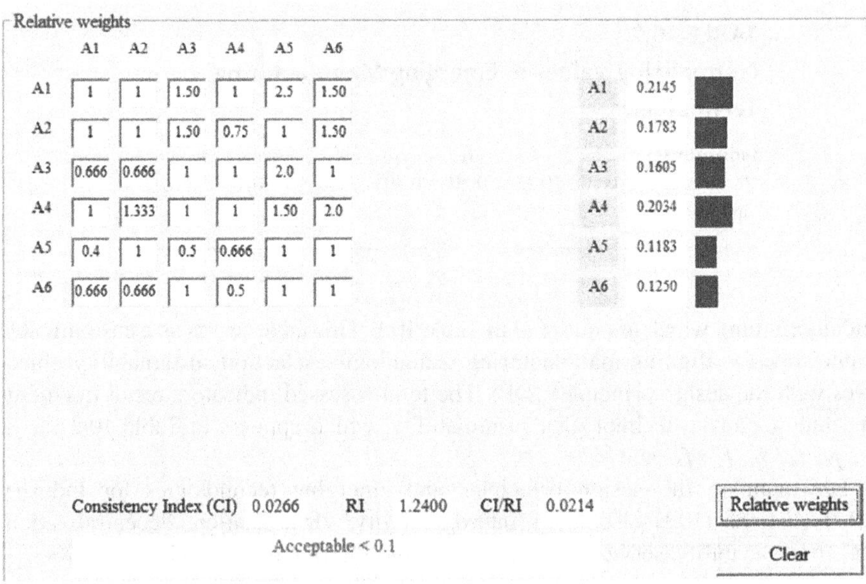

FIGURE 10.5 Relative weights between design principles in Industry 4.0.

Then, by using Equation (10.8) and substituting the values of relative weights coming from Figure 10.5, the sustainability index of technologies selection (TS) at any time t, $s_{TS}(t)$ is evaluated by utilizing Equation (10.8), and the result is 0.3534 (35.34%).

$$S_{TS}(t) = 0.2145\ g_1(t) + 0.1783\ g_2(t) + 0.1605\ g_3(t) + 0.2034\ g_4(t) + 0.1183\ g_5(t) + 0.1250\ g_6(t)$$

$$S_{TS}(t) = 0.2145(0.4519) + 0.1783(0.50) + 0.1605\ (0.4433) + 0.2034\ (0.4575) + 0.1183\ (0.4660) + 0.1250\ (0.4575) = 0.3534$$

10.6 CONCLUDING REMARKS

The second layer of Industry 4.0 sustainability, based on an architecture perspective, was proposed and analyzed through the chosen layer, referred to as the process of technology selection for Industry 4.0. This chosen layer builds upon the six design principles (DP) and eight major customized emerging technologies (ET), providing a framework for implementing and sustaining Industry 4.0.

Industry 4.0 is not a standalone invention but rather an evolved version of Industry 3.0. This evolution has been driven by the reduced costs of enabling emerging technologies (digital and manufacturing), which have been customized to Industry 4.0 requirements. The integration of these technologies with the internet has facilitated enhanced connectivity and collaboration, creating the foundation for intelligent manufacturing systems.

The six design principles (DP) and advanced emerging technologies play a pivotal role in the implementation and sustainability of Industry 4.0. Each principle exhibits unique characteristics and values, while each technology possesses distinct specifications and attributes. These elements are dynamic, evolving, and varying between plants or factories. The ability to adapt these principles and technologies to specific contexts ensures the long-term viability and effectiveness of smart manufacturing systems.

The mapping between design principles and emerging technologies has been presented both mathematically and logically, offering a structured approach to integration. The sustainability index for Industry 4.0 technological selection has been assessed to evaluate the alignment of technologies and design principles with sustainability objectives. This index provides a quantitative measure of the effectiveness of the chosen layer in achieving sustainable smart manufacturing.

The architecture perspective system forms a critical link between the chosen layer, the preceding core layer, and the subsequently given layer, as illustrated in Figure 10.6. This connection underscores the importance of mental frameworks and decision-making processes in aligning technological selection with organizational goals and sustainability objectives.

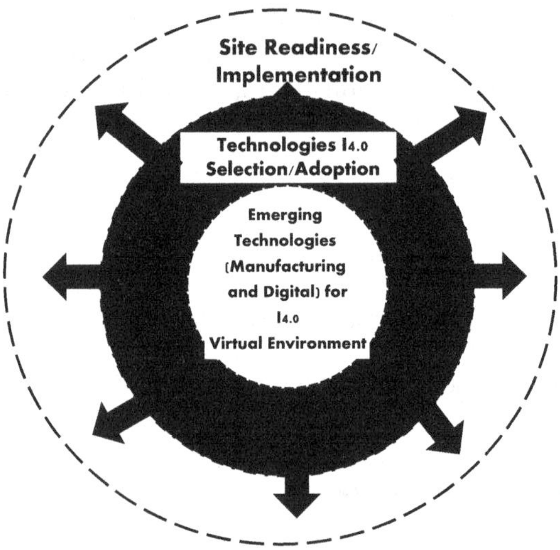

FIGURE 10.6 Affecting chosen layer (technologies selection) on the given layer.

REFERENCES

Garbie, I., (2013), "DFSME: Design for Sustainable Manufacturing Enterprises (An Economic Viewpoint)", *International Journal of Production Research*, Vol. 51, No. 2, pp. 479–503.

Garbie, I., (2014), "An Analytical Technique to Model and Assess Sustainable Development Index in Manufacturing Enterprises", *International Journal of Production Research*, Vol. 52, No. 16, pp. 4876–4915.

Garbie, I., and Garbie, A., (2020a), "Outlook of Requirements of Manufacturing Systems for 4.0", *3rd International Conference of Advances in Science and Engineering Technology (Multi-Conferences) ASET 2020*, February 4–6, 2020, Dubai, UAE.

Garbie, I., and Garbie, A., (2020b), "A New Analysis and Investigation of Sustainable Manufacturing through a Perspective Approach", *3rd International Conference of Advances in Science and Engineering Technology (Multi-Conferences) ASET 2020*, February 4–6, 2020, Dubai, UAE.

Garbie, I., and Garbie, A., (2020c), "Sustainability and Manufacturing: A Conceptual Approach", *Proceedings of the Industrial and Systems Engineering Research Conference (IISE Annual Conference and Expo 2020)*, November 1–3, 2020, New Orleans, LA, USA, 6 pp.

Garbie, I., and Garbie, A., (2022), "Identifying and Estimating the Implementation Level for Industry 4.0", *The 4th International Conference of Advances in Science and Engineering Technology (Multi-Conferences) ASET 2022*, February 21–24, 2022, Dubai, UAE, 6 pp.

Garbie, I., and Parsaei, H. (2022), *Reconfigurable Manufacturing Enterprises for Industry 4.0*, 1st Edition. CRC Press-Taylor & Francis, Boca Raton, FL, USA.

Garbie, I., Parsaei, H. R., and Leep, H. R., (2008a), "Measurement of Needed Reconfiguration Level for Manufacturing Firms", *International Journal of Agile Systems and Management*, Vol. 3, Nos. 1/2, pp. 78–92 (Special Issue on Complexity in Manufacturing).

Garbie, I., Parsaei, H. R., and Leep, H. R., (2008b), "A Novel Approach for Measuring Agility in Manufacturing Firms", *International Journal of Computer Applications in Technology*, Vol. 32, No. 2, pp. 95–103.

Grilchrist, A., (2016), *Industry 4.0: the Industrial Internet of Things*. Apress, Bangken, Thailand.

Lins, T., and Oliveira, R. A. R., (2020), "Cyber-Physical Production Systems Retrofitting in Context of Industry 4.0", *Computers and Industrial Engineering*, Vol. 139, 106193.

Ooi, L. L., Teh, S. Y., and Cheang, P. Y. S., (2023), "The Impact of Lean Production on Sustainable Organizational Performance: The Moderating Effect of Industry 4.0 Technologies Adoption", *Management Research Review*, Vol. 46 No. 12, pp. 1802–1836.

Pawar, K., Kasat, K., Deshpande, A. P., and Shaikh, N., (2023), "Adoption of Industry 4.0 Technologies for Performance Management", *1st DMIHER International Conference on Artificial Intelligence in Education and Industry 4.0*, Wardha, Maharashtra, India, pp. 1–5.

Qin, S.-F., and Cheng, K., (2017), "Future Digital Design and Manufacturing: Embracing Industry 4.0 and Beyond", *Chinese Journal of Mechanical Engineering*, Vol. 30, pp.1047–1049.

Tortorella, G. L., Saurin, T. A., Fogliatto, F. S., Mendoza, D. T., Moyano-Fuentes, J., Gaiardelli, P., Seyedghorban, Z., Vassolo, R., Vergara, A. F. M. C., Sunder M. V. Sreedharan, R., Sena, A. A., Forstner, F. F., and De Anda, E. M., (2024), "Digitalization of Maintenance: Exploratory Study on the Adoption of Industry 4.0 Technologies and Total Productive Maintenance Practices", *Production Planning and Control*, Vol. 35, No. 4, pp. 352–372.

Zaidi, N. I. N., Nazmudeen, M. S., and Mohiddin, F., (2022), "Identifying Factors Contributing to the Level of Industry 4.0 Technologies Adoption among SMEs in Different Countries", *International Journal of Engineering and Technology*, Vol. 14, No. 3, pp. 43–51.

Zhou, B., and Zheng, L., (2023), "Technology-Pushed, Market-Pulled, or Government-Driven? The Adoption of Industry 4.0 Technologies in a Developing Economy", *Journal of Manufacturing Technology Management (JMTM)*, Vol. 34, No. 9, pp. 115–138.

11 Sustainable Site Readiness/Implementation

11.1 INTRODUCTION TO SITE READINESS/IMPLEMENTATION

The implementation of Industry 4.0 in manufacturing industries poses significant challenges, particularly from a psychological perspective. Industry 4.0 represents the adoption of advanced manufacturing technologies, including cyber-physical systems (CPS), the Internet of Things (IoT), virtual and augmented reality (VR/AR), cloud computing and manufacturing (CC/CM), big data analytics (BDA), additive manufacturing (AM), and cybersecurity (CS).

Several factors act as barriers, drivers, and challenges to Industry 4.0 implementation. These include technical challenges (TEC), risks (RIS), critical success factors (CSF), opportunities OPP), and enablers of implementation (ENA) (Garbie and Parsaei, 2022; Garbie and Garbie, 2022). Although these analyses are exploratory and qualitative in nature, they serve as valuable guides for evaluating the level of implementation.

Manufacturing systems and enterprises often face managerial constraints that can hinder the adoption of Industry 4.0. These constraints include localized management practices, short-term strategic planning, insufficient expertise, dysfunctional organizational structures, limited resources, and the absence of appropriate methods and techniques (Garbie and Garbie, 2022; Garbie et al., 2023).

Achieving sustainability in Industry 4.0 or smart manufacturing can be seen as the "third stream" in the broader context of sustainability efforts. The objective of this stream is to demonstrate that Industry 4.0 can not only be implemented but can also remain sustainable over time. This stream has a psychological connection to the preceding "core" and "chosen" streams (see Figure 11.1). Figure 11.1 depicts the chosen stream as a bridge between the core and third streams. The overarching goal of this chapter is to identify and analyze the motivations, drivers, barriers, and challenges of implementing Industry 4.0 while maintaining sustainable site readiness.

11.2 ISSUES OF SITE READINESS

Although Industry 4.0 emerged as a transformative concept approximately a decade ago, many manufacturing industries still lack a clear vision for its implementation. Exploring the sustainability of Industry 4.0 at various implementation levels has become an exciting and essential area of study. To establish a framework that clarifies the vision for implementing Industry 4.0, Figure 11.2 provides a conceptual illustration. This framework is guided by addressing five critical "What" questions (5WQs), which are essential to evaluating site readiness and implementation sustainability:

DOI: 10.1201/9781003644835-15

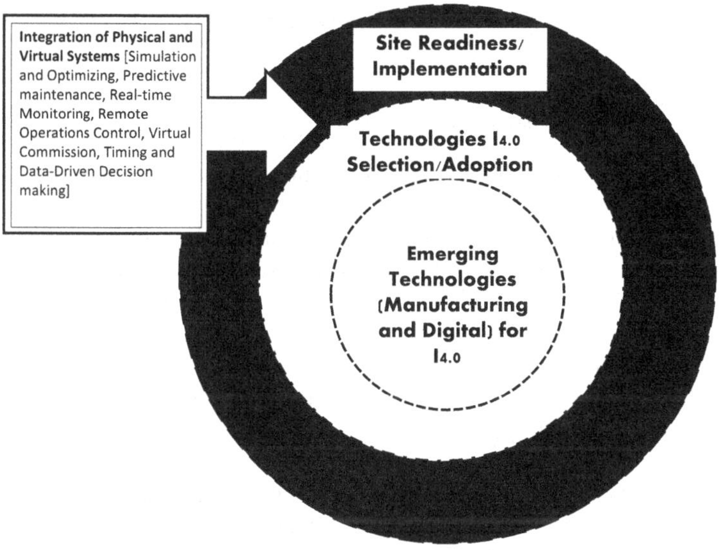

FIGURE 11.1 Site readiness for I4.0 concerning other layers.

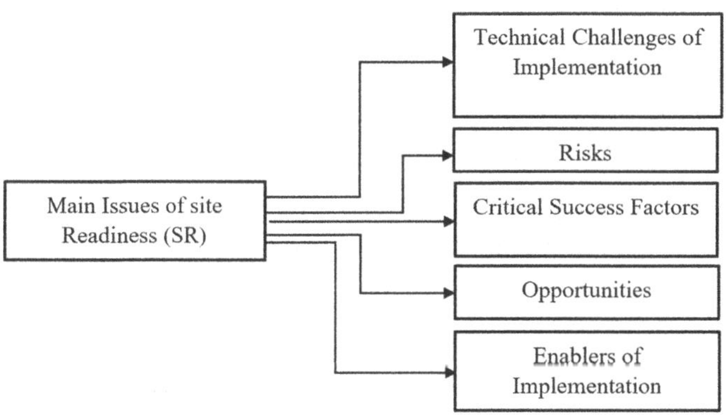

FIGURE 11.2 Main issues of site readiness.

1. What are the technical challenges facing Industry 4.0?
2. What are the risks involved in implementing Industry 4.0?
3. What are the critical success factors required for successful implementation of Industry 4.0?
4. What are the opportunities available for Industry 4.0 implementation?
5. What are the enablers necessary to assess the implementation status of Industry 4.0?

By systematically addressing these questions, this chapter provides a structured approach to evaluate the barriers, risks, and opportunities associated with implementing Industry 4.0, while highlighting key factors that enable and sustain its adoption.

11.2.1 Technical Challenges of Implementation

Technical challenges (h_1) represent one of the most critical aspects of implementing Industry 4.0.

Smart hardware and software are the foundational pillars for constructing smart manufacturing systems or enterprises. These technologies include:

- smart controllers for machines,
- smart industrial wireless networks (IWN),
- big data analytics (BDA) software for manufacturing, and
- integrated information systems.

The identification of technical challenges (h_1) has been supported by various studies, which highlight several key elements. Recommendations from Wang et al. (2016), Moktadir et al. (2018), Garbie and Parsaei (2022), and Garbie and Garbie (2024a,b) provide a comprehensive framework for addressing these challenges by taking information technology (IT), operation technology (OT), big data analytical (BDA), and cyber security (CS) into consideration:

- Design of facilities and layout (h_{11})
- Adopting Modular artifacts (h_{12})
- High-speed integration with the (IT) and (OT) (h_{13})
- Applying the BDA to manufacturing (h_{14})
- Protecting Data and the CS (h_{15})
- Initiation Smart decision-making (h_{16})
- Investment cost (h_{17})
- Deficiency of infrastructure (technologies and/or physical) (h_{18})
- Lack of connection among partnerships (h_{19})
- Less commitment to Industry 4.0 from the upper management level (h_{110})
- Deficiency of adopting reconfiguration strategy for manufacturing companies (h_{111})
- Level of professional employees for implementing Industry 4.0 (h_{112})
- Social side-effects of implementing Industry 4.0 (h_{113})
- Biological side effects of implementing Industry 4.0 (h_{114})

To evaluate the existing value of h_1, E^{h_1}, the following questionnaire (Table 11.1) will be distributed, and the collected values from the manufacturing organizations will be presented and calculated as in the following Equation (11.1).

$$E^{h_1} = \frac{\sum_{h_{11}}^{h_{114}} \text{value of } \left(h_{11} + h_{12} + \cdots + h_{114}\right)}{70} \tag{11.1}$$

TABLE 11.1
The Technical Challenges (h_1) Questionnaire

Indicator	Relevance Performance	1	2	3	4	5
h_{11}	Possibility of system modeling and analysis for adopting I4.0					
h_{12}	Possibility of using modular and flexible physical artifacts					
h_{13}	Possibility of integrating IT and OT with high-speed internet					
h_{14}	Possibility of using BDA for manufacturing data					
h_{15}	Possibility of adopting the CS					
h_{16}	Possibility of using smart decision-making mechanism					
h_{17}	Possibility of investment high cost for Industry 4.0					
h_{18}	Possibility to invest in infrastructure (technological and physical)					
h_{19}	Possibility to connect among manufacturing companies					
h_{110}	Possibility to adopt a clear strategy toward Industry 4.0					
h_{111}	Possibility to reconfigure manufacturing enterprises for Industry 4.0					
h_{112}	Possibility to acquire highly qualified personal					
h_{113}	Possibility to minimize unemployment and social side-effects					
h_{114}	Possibility to minimize environmental side effects					

Note: 1 means the lowest possibility; 5 means the highest possibility.

11.2.2 Risks

The adoption of Industry 4.0 is accompanied by significant risks (h_2), primarily due to a lack of expertise and the prevalence of short-term strategic thinking. These risks (RIS) have been extensively identified and analyzed in the literature. Recommendations from Moktadir et al. (2018), Bai et al. (2020), and Garbie and Parsaei (2022) provide valuable insights into understanding and mitigating these risks falling into the areas below:

- Lack of professional employees (h_{21})
- Adopting Short-term policy and strategy (h_{22})
- Deficiency in technology investment (h_{23})

TABLE 11.2
The Risks (h_2) Questionnaire

Indicator	Relevance Performance	1	2	3	4	5
h_{21}	(%) or level of lack of professional employees					
h_{22}	(%) or level of long-term strategy					
h_{23}	(%) or level of investment in advanced manufacturing technologies					
h_{24}	(%) or level of employees afraid of surveillance of their work					
h_{25}	(%) of unemployment					
h_{26}	(%) or level of organizational resistance					
h_{27}	(%) or level of equipment and technology integration					
h_{28}	(%) or level of maturity for implementation					
h_{29}	(%) or level of support from the government					
h_{210}	(%) or level of manufacturing process complexity					

Note: 1 means the lowest possibility; 5 means the highest possibility.

- Very close attention and monitoring of employees (h_{24})
- Unemployment as a social side-effect (h_{25})
- Lack from the middle and senior management levels toward implementing Industry 4.0 (h_{26})
- Integration between current machines and Industry 4.0 technologies (h_{27})
- Lack of strategies and maturity for implementation (h_{28})
- Lack of support from governmental agencies (h_{29})
- Manufacturing process complexity (h_{210})

To estimate the existing value of h_2, E^{h_2}, the following questionnaire (Table 11.2) will be distributed, and the collected values from the manufacturing organizations will be presented and calculated as in the following Equation (11.2).

$$E^{h_2} = \frac{\sum_{h_{21}}^{h_{210}} \text{value of } \left(h_{21} + h_{22} + \cdots + h_{210}\right)}{50} \tag{11.2}$$

11.2.3 Critical Success Factors

Critical success factors (h_3) represent the third key issue in implementing Industry 4.0, alongside technical challenges (h_1) and risks (h_2). For example, training is identified as one of the most crucial success factors, while the reconfiguration of manufacturing systems is recognized as a significant and unique opportunity offered by Industry 4.0 (Garbie and Garbie, 2020a,b,c).

The critical success factors (h_3) have been identified based on practical experience and are informed by the following key points, as recommended by Moeuf et al. (2019) and Garbie and Parsaei (2022):

- **Strong Commitment from Upper Leadership for Conducting Industry 4.0** (h_{31}): Effective leadership is essential to drive the adoption and integration of Industry 4.0.
- **A Clear Strategy for Communication form for Industry 4.0** (h_{32}): Ensuring transparency and shared understanding among stakeholders is crucial.
- **Enhance Competencies and Training of Employees** (h_{33}): Continuous training initiatives help employees acquire the necessary skills and knowledge for Industry 4.0.
- **Conducting Pilot Test before Implementing Industry 4.0** (h_{34}): Comprehensive planning and analysis mitigate risks and support informed decision-making.
- **Adopting Strategy for Continuous Improvement** (h_{35}): Regularly refining processes ensure long-term success and adaptability.
- **Fully Utilization of Available Company Resources (e.g., Data)** (h_{36}): Leveraging data facilitates decision-making and process optimization.
- **Adopting Vertical Integration** (h_{37}): Ensuring coordination and integration across all levels of the organization is vital.
- **Industrial Organization Flexibility Level** (h_{38}): Adaptive and agile management practices enable efficient responses to challenges and opportunities.

To evaluate the existing value of h_3, E^{h_3}, the following questionnaire (Table 11.3) will be distributed, and the collected values from the industrial organizations will be presented and calculated as in the following Equation (11.3).

TABLE 11.3
The Critical Success Factors (h_3) Questionnaire

Indicator	Relevance Performance	1	2	3	4	5
h_{31}	Possibility of getting professional employees for Industry 4.0					
h_{32}	Possibility of using a clear strategy for Industry 4.0					
h_{33}	Possibility of enhancing employee training					
h_{34}	Possibility of conducting a pilot test before implementing Industry 4.0					
h_{35}	Possibility of adopting a continuous improvement strategy					
h_{36}	Possibility of utilizing manufacturing company resources (e.g., data)					
h_{37}	Possibility of alignment along vertical integration					
h_{38}	Possibility of industrial organizations' management flexibility level					

Note: 1 means the lowest possibility; 5 means the highest possibility.

$$E^{h_3} = \frac{\sum_{h_{31}}^{h_{310}} \text{value of } (h_{31} + h_{32} + \cdots + h_{38})}{40} \quad (11.3)$$

11.2.4 Opportunities

Garbie and Parsaei (2022) identified several key opportunities (h_4) for the successful implementation of Industry 4.0. These opportunities include:

- Strong backing from leadership ensures alignment with organizational goals and provides necessary resources (h_{41}).
- Leveraging advanced data analytics and simulations enhances decision-making and predictive capabilities (h_{42}).
- Adopting Industry 4.0 technologies optimizes processes, boosting efficiency, and productivity (h_{43}).
- Updating business models to integrate smart technologies enables greater adaptability and innovation (h_{44}).
- Industry 4.0 technologies improve market positioning by fostering efficiency, innovation, and customer satisfaction (h_{45}).

To evaluate the existing value of h_4, E^{h_4}, the following questionnaire (Table 11.4) will be distributed, and the collected values from the manufacturing organizations will be presented and calculated as in the following Equation (11.4).

$$E^{h_4} = \frac{\sum_{h_{41}}^{h_{45}} \text{value of } (h_{41} + h_{42} + \cdots + h_{45})}{25} \quad (11.4)$$

TABLE 11.4
The Opportunities (h_4) Questionnaire

Indicator	Relevance Performance	1	2	3	4	5
h_{41}	Possibility of strong backing and support from leadership					
h_{42}	Possibility of leveraging advanced data analytics and simulations					
h_{43}	Possibility of optimizing processes and performance					
h_{44}	Possibility of updating and upgrading business models					
h_{45}	Possibility of improving marketing					

Note: 1 means the lowest possibility; 5 means the highest possibility.

11.2.5 Enablers of Implementation

The enablers of implementation (h_5) constitute the final issue in achieving sustainability for site readiness in Industry 4.0. These enablers primarily focus on enabling technologies, which are categorized into two main types: base technologies and front-end technologies (Frank et al., 2019).

- **Base Technologies**: These foundational technologies include big data analytics (BDA), cloud computing/cloud manufacturing (CC/CM), additive manufacturing (AM), wireless-networked manufacturing (WNM), horizontal and vertical integration (HVI), the Internet of Things (IoT), and cyber-physical production systems (CPPS). These technologies, previously discussed as part of the core stream in Chapter 9, are integral to the successful implementation of Industry 4.0.
- **Front-End Technologies**: In addition to base technologies, front-end technologies address areas such as smart product life cycle management, collaborative productivity, top-level management, mass customization, smart supply chain management, professional training and development, and operational efficiency (see Figure 11.3). These technologies complement the base technologies and enhance their application in Industry 4.0 contexts.

The critical technological enablers for implementing Industry 4.0 are summarized in this section, as highlighted by Garbie and Parsaei (2022). Based on the work of Sivananda et al. (2020), 14 key enablers have been identified:

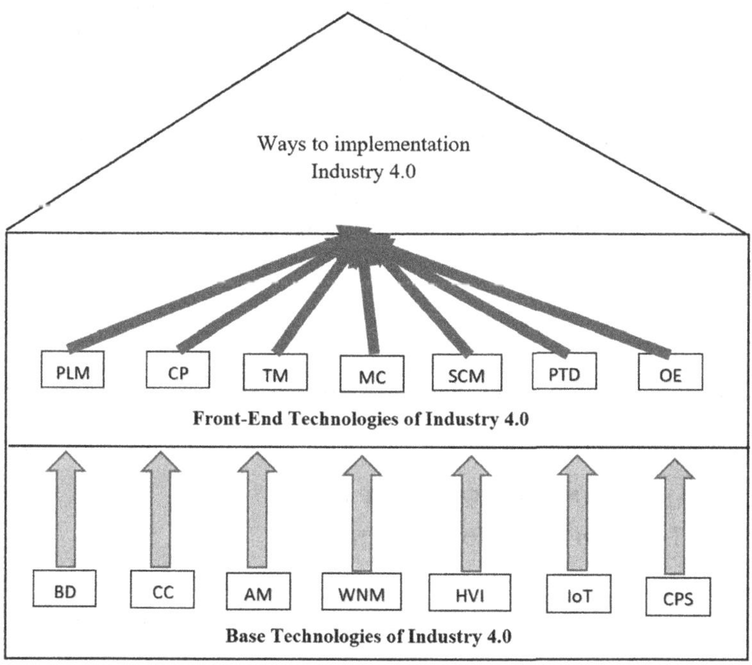

FIGURE 11.3 Technologies for Industry 4.0 implementation (Garbie and Parsaei, 2022).

- **Applied Big Data Analytical (BDA)** (h_{51}): Advanced analytics for handling large datasets.
- **Working under the Umbrella of Cloud Computing (CC) or Cloud Manufacturing (CM)** (h_{52}): Cloud-based solutions for seamless data storage and processing.
- **Using Additive Manufacturing (AM) Technologies** (h_{53}): Advanced production techniques such as 3D printing.
- **Adoption of Wireless-Networked (WN) Systems** (h_{54}): Integration of wireless technologies in manufacturing environments.
- **Integration between Horizontal and Vertical (HV) Communication** (h_{55}): Alignment across operational and strategic levels.
- **Using the Industrial Internet of Things (IIoT)** (h_{56}): Connected devices enabling data exchange and automation.
- **Incorporating Cyber-Physical Production Systems (CPPSs)** (h_{57}): Integration of physical and digital systems.
- **Optimizing Product Life Cycle (PLC) Management (PLM)** (h_{58}): Optimizing the management of product development and lifecycle.
- **Perspective of Collaborative and Enhancement Productivity** (CEP) (h_{59}): Enhancing teamwork and shared goals.
- **Motivation and Encouragement from Top-Level Management (TLM)** (h_{510}): Strategic leadership for Industry 4.0 initiatives.
- **Follow Mass Customization Strategy (MC)** (h_{511}): Tailoring production to meet individual customer needs.
- **Follow the Procedures of Smart Supply Chain Management (SSCM)** (h_{512}): Improving supply chain efficiency and responsiveness.
- **Motivation for Professional Development and Training (PDT)** (h_{513}): Equipping employees with Industry 4.0 skills.
- **Monitoring and Follow Up Operational Efficiency (OE)** (h_{514}): Streamlining processes for improved productivity and resource utilization.

To estimate the existing value of h_5, E^{h_5}, the following questionnaire (Table 11.5) will be distributed, and the collected values from the manufacturing organizations will be presented and calculated as in the following Equation (11.5).

$$E^{h_5} = \frac{\sum_{h_{51}}^{h_{514}} \text{value of } \left(h_{51} + h_{52} + \cdots + h_{514}\right)}{70} \tag{11.5}$$

11.3 ANALYSIS AND ASSESSMENT OF SUSTAINABLE SITE READINESS

The "given stream" of Industry 4.0 or smart manufacturing refers to site readiness (SR), which is developed as a progression from the previous streams: "core" (ET) and "chosen" (TS). This stream focuses on ensuring the plant or factory is both physically and virtually prepared to implement Industry 4.0 technologies effectively.

TABLE 11.5
The Enablers of Implementation (h_5) Questionnaire

Indicator	Relevance Performance	1	2	3	4	5
h_{51}	Possibility of using the big data analytical (BDA)					
h_{52}	Possibility of cloud computing (CC)/ cloud manufacturing (CM)					
h_{53}	Possibility of adopting additive manufacturing (AM)					
h_{54}	Possibility of adopting wireless-networked manufacturing					
h_{55}	Possibility of conducting integration between horizontal and vertical					
h_{56}	Possibility of adopting the Internet of Things					
h_{57}	Possibility of installing a Cyber-physical system					
h_{58}	Possibility of analyzing product life cycle management					
h_{59}	Possibility of conducting collaborative productivity					
h_{510}	Encouragement from Top-level management					
h_{511}	Encouragement to follow up mass customization philosophy					
h_{512}	Possibility of implementing smart supply chain management					
h_{513}	Possibility to enhance professional training and development					
h_{514}	Possibility to improve operational efficiency					

Note: 1 means the lowest possibility; 5 means the highest possibility.

The outcome of this stream is reflected in various performance measures that integrate physical and virtual systems, making the site feasible for Industry 4.0 adoption.

Key performance measures include:

- simulation and optimization,
- predictive maintenance,
- real-time monitoring,
- remote operations control,
- virtual commissioning,
- timing improvements, and
- data-driven decision-making.

The sustainability level of Industry 4.0/Smart Manufacturing in terms of site readiness at any time t, $S_{SR}(t)$, can be mathematically expressed based on the type of issues of site readiness. These issues include:

- technical challenges of implementation (h_1),
- risks (h_2),
- critical success factors (h_3),
- opportunities (h_4), and
- enablers of implementation(h_5).

The relationship is formalized in Equation (11.6), which models the sustainability index of the site readiness by incorporating these key factors.

$$S_{SR}(t) = f(h_1, h_2, h_3, h_4, h_5) = \left\{ \begin{array}{c} h_1 \\ h_2 \\ h_3 \\ h_4 \\ h_5 \end{array} \right\} = \sum_{i=1}^{5} h_i(t)/5 \tag{11.6}$$

where:

$h_i(t)$ in Equation (11.6) can be calculated using the following Equation (11.7)

$$h_i(t) = \frac{h_i(t)\text{current} - h_i(t)\min}{h_i(t)\max - h_i(t)\min} \tag{11.7}$$

where:

$h_i(t)$ is the normalized value of the integration of physical and virtual systems for site readiness (SR) at any given time t,

$h_i(t)$ current is the existing value of the integration of physical and virtual systems for site readiness (SR) at any time t,

$h_i(t)\max$ and $h_i(t)\min$ are the maximum and minimum values of the integration of physical and virtual systems for site readiness (SR) at any time t, respectively.

11.3.1 For Technical Challenges of Implementation (H_1)

$h_1(t)$ = normalizing value of technical challenges of implementation (h_1) at any time t.

$$h_1(t) = \frac{E^{h_1} - I_{\min}^{h_1}}{I_{\max}^{h_1} - I_{\min}^{h_1}} \tag{11.8}$$

E^{h_1} = the existing value of technical challenges of implementation (h_1) at any time t and this value was measured from the above questionnaire (Table 11.1) and Equation (11.1),

$I_{\max}^{h_1}$ = maximum value of technical challenges of implementation (h_1) at any time t,

$I_{\min}^{h_1}$ = minimum value of technical challenges of implementation (h_1) at any time t.

11.3.2 For Risks (H_2)

$h_2(t)$ = normalizing value of risks (h_2) at any time t.

$$h_2(t) = \frac{E^{h_2} - I_{\min}^{h_2}}{I_{\max}^{h_2} - I_{\min}^{h_2}} \tag{11.9}$$

E^{h_2} = the existing value of risks (h_2) at any time t and this value was measured from the above questionnaire (Table 11.2) and Equation (11.2),

$I_{\max}^{h_2}$ = maximum value of risks (h_2) at any given time t,

$I_{\min}^{h_2}$ = minimum value of risks (h_2) at any given time t.

11.3.3 For Critical Success Factors (H_3)

$h_3(t)$ = normalizing value of critical success factors (h_3) at any time t.

$$h_3(t) = \frac{E^{h_3} - I_{\min}^{h_3}}{I_{\max}^{h_3} - I_{\min}^{h_3}} \tag{11.10}$$

E^{h_3} = the existing value of critical success factors (h_3) at any time t and this value was measured from the above questionnaire (Table 11.3) and Equation (11.3)

$I_{\max}^{h_3}$ = maximum value of critical success factors (h_3) at any given time t,

$I_{\min}^{h_3}$ = minimum value of critical success factors (h_3) at any given time t.

11.3.4 For Opportunities (H_4)

$h_4(t)$ = normalizing value of opportunities (h_4) at any given time t.

$$h_4(t) = \frac{E^{h_4} - I_{\min}^{h_4}}{I_{\max}^{h_4} - I_{\min}^{h_4}} \tag{11.11}$$

E^{h_4} = the existing value of opportunities (h_4) at any time t and this value was measured from the above questionnaire (Table 11.4) and Equation (11.4)

$I_{\max}^{h_4}$ = maximum value of opportunities (h_4) at any time t,

$I_{\min}^{h_4}$ = minimum value of opportunities (h_4) at any time t.

11.3.5 For Enablers of Implementation (H_5)

$h_5(t)$ = normalizing value of enablers of implementation (h_5) at any time t.

$$h_5(t) = \frac{E^{h_5} - I_{\min}^{h_5}}{I_{\max}^{h_5} - I_{\min}^{h_5}} \tag{11.12}$$

E^{h_5} = the existing value of enablers of implementation (h_5) at any time t and this value was measured from the above questionnaire (Table 11.5) and Equation (11.5),

$I^{h_5}_{max}$ = maximum value of enablers of implementation (h_5) at any time t,
$I^{h_5}_{min}$ = minimum value of enablers of implementation (h_5) at any time t.

11.4 NUMERICAL EXAMPLE 11.1

This illustrative example demonstrates the proposed framework for assessing the sustainability index of Industry 4.0 (I4.0) for site readiness (SR) issues. The data used for this analysis pertain to the following factors:

The data used for this analysis pertain to the following factors:

- technical challenges (h_1),
- risks (h_2),
- critical success factors (h_3),
- use of opportunities (h_4), and
- enablers of implementation (h_5).

Relevant data for these factors are provided in Tables 11.6–11.10. For technical challenges (h_1), the existing current value of h_1, E^{h_1}, is estimated using Equation (11.1) and

TABLE 11.6
Data for Technical Challenges (h_1)

Indicator	Relevance Performance	1	2	3	4	5
h_{11}	Possibility of system modeling and analysis for adopting I4.0				*	
h_{12}	Possibility of using modular and flexible physical artifacts				*	
h_{13}	Possibility of using high-speed integration with IT and OT			*		
h_{14}	Possibility of using specific big data analytics in manufacturing			*		
h_{15}	Possibility of adopting cyber security				*	
h_{16}	Possibility of smart decision-making mechanism				*	
h_{17}	Possibility of investment for Industry 4.0					*
h_{18}	Possibility to invest in technological infrastructure					*
h_{19}	Possibility to connect among manufacturing companies					*
h_{110}	Possibility to adopt a clear strategy toward Industry 4.0					*
h_{111}	Possibility to reconfigure manufacturing enterprises for Industry 4.0			*		
h_{112}	Possibility to acquire highly qualified personal				*	
h_{113}	Possibility to minimize unemployment and social side-effects				*	
h_{114}	Possibility to minimize environmental side effects				*	

TABLE 11.7
Data for the Risks (h_2)

Indicator	Relevance Performance	1	2	3	4	5
h_{21}	(%) or level of professional skills of employees			*		
h_{22}	(%) or level of long-term strategy				*	
h_{23}	(%) or level of investment in advanced manufacturing technologies					*
h_{24}	(%) or level of employees afraid of surveillance of their work				*	
h_{25}	(%) of unemployment		*			
h_{26}	(%) or level of organizational resistance			*		
h_{27}	(%) or level of equipment and technology integration				*	
h_{28}	(%) or level of maturity for implementation			*		
h_{29}	(%) or level of support from the government			*		
h_{210}	(%) or level of complexity in the manufacturing/ production process			*		

TABLE 11.8
Data for the Critical Success Factors (h_3)

Indicator	Relevance Performance	1	2	3	4	5
h_{31}	Possibility of getting professional employees for Industry 4.0			*		
h_{32}	Possibility of using a clear strategy for Industry 4.0				*	
h_{33}	Possibility of enhancing employee training					*
h_{34}	Possibility of conducting a pilot test before implementing Industry 4.0				*	
h_{35}	Possibility of adopting a continuous improvement strategy					*
h_{36}	Possibility of utilizing manufacturing company resources (e.g., data)					*
h_{37}	Possibility of alignment along vertical integration					*
h_{38}	Possibility of industrial organizations' management flexibility level			*		

data in Table (11.6) as follows. This calculation indicates that the current value of h_1 is higher than the medium level, suggesting significant technical challenges present in the current site readiness scenario.

TABLE 11.9
Data for the Opportunities (h_4)

Indicator	Relevance Performance	1	2	3	4	5
h_{41}	Possibility of strong backing and support from leadership					*
h_{42}	Possibility of leveraging advanced data analytics and simulations					*
h_{43}	Possibility of optimizing processes and performance					*
h_{44}	Possibility of updating and upgrading business models				*	
h_{45}	Possibility of improving marketing				*	

TABLE 11.10
Data for the Enablers of Implementation (h_5)

Indicator	Relevance Performance	1	2	3	4	5
h_{51}	Possibility of using big data Analytics		*			
h_{52}	Possibility of using cloud computing (CC)/ cloud manufacturing (CM)		*			
h_{53}	Possibility of adopting additive manufacturing (AM)				*	
h_{54}	Possibility of adopting wireless-networked manufacturing				*	
h_{55}	Possibility of integrating between horizontal and vertical					*
h_{56}	Possibility of adopting the Internet of Things				*	
h_{57}	Possibility of installing a Cyber-physical system					*
h_{58}	Possibility of analyzing product life cycle management					*
h_{59}	Possibility of conducting collaborative productivity					*
h_{510}	Encouragement from Top-level management					*
h_{511}	Encouragement to follow up mass customization philosophy				*	
h_{512}	Possibility of implementing smart supply chain management				*	
h_{513}	Possibility to enhance professional development and training					*
h_{514}	Possibility to improve operational efficiency				*	

$$E^{h_1} = \frac{3\times3+4\times7+5\times4}{70} = 0.8142$$

For risks (h_2), the existing current value of h_2, E^{h_2}, is estimated using Equation (11.2) data in Table (11.7) as follows. It seems that the existing value of risks is medium. The calculation indicates that the current value of h_2 is at a medium level, suggesting that while risks are present, they are not excessively high and may be manageable within the site readiness framework.

$$E^{h_2} = \frac{1\times2+2\times3+3\times4+1\times5}{50} = 0.60$$

For critical success factors (h_3), the existing current value of h_3, E^{h_3}, is estimated using Equation (11.3) and data in Table (11.8) as follows. It seems that the existing value of critical success factors is higher than the medium and very close to the highest. This suggests that site readiness has strong critical success factors likely to significantly contribute to the successful implementation of Industry 4.0

$$E^{h_3} = \frac{2\times3+2\times4+4\times5}{40} = 0.85$$

For opportunities (h_4), the existing current value of h_4, E^{h_4}, is estimated using Equation (11.4) and data in Table (11.9) as follows. It seems that the existing value of opportunities is the highest. This suggests that the site readiness has maximized the available opportunities, providing a strong foundation for the successful implementation of Industry 4.0.

$$E^{h_4} = \frac{2\times4+3\times5}{25} = 0.92$$

For enablers of implementation (h_5), the existing current value of h_5, E^{h_5}, is estimated using Equation (11.5) and data in Table (11.10) as follows. It seems that the existing value of enablers of implementation is higher than the medium and close to the highest.

$$E^{h_5} = \frac{2\times2+4\times6+5\times6}{70} = 0.8285$$

The total site readiness issues result appears in Table 11.11. The sustainability index of site readiness (*SR*) at any time t, $s_{SR}(t)$ is evaluated by utilizing Equation (11.6), and the results are outlined in Table 11.11 as 80.25%. This value appears high.

TABLE 11.11
Sustainability Index in Site Readiness

Indicator (h)	h_1	h_2	h_3	h_4	h_5
Index of Issues	0.8142	0.6000	0.8500	0.9200	0.8285
$S_{SR}(t)$	0.8025				

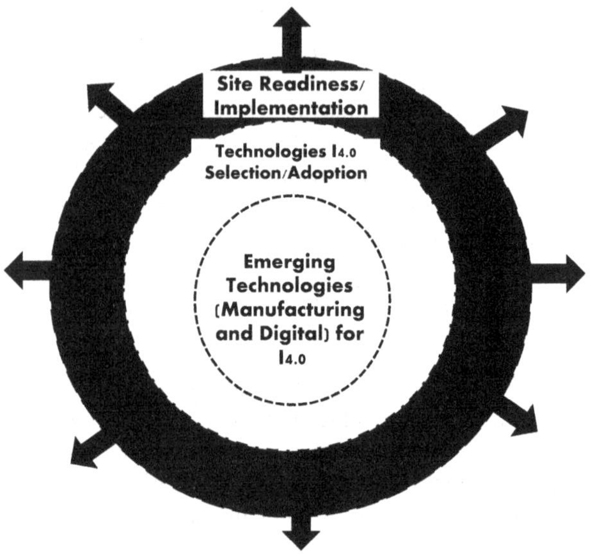

FIGURE 11.4 Outcomes from the given layer (site readiness) based on the previous layers.

11.5 CONCLUDING REMARKS

The third layer of sustainable Industry 4.0/smart manufacturing, viewed from an architecture perspective, was proposed and analyzed through the "given layer", also referred to as site readiness for Industry 4.0 (I4.0). The framework of the given layer is primarily based on five key issues of site readiness: technical challenges, risks, critical success factors, opportunities, and enablers of implementation.

These issues were systematically analyzed and presented within the context of Industry 4.0/smart manufacturing using a survey questionnaire framework. The existing values for these issues were mathematically and logically estimated and explained. Additionally, the sustainability index of site readiness was proposed, illustrated, and assessed through a numerical example.

The architectural system plays a crucial role in Industry 4.0/smart manufacturing by bridging the given layer with the preceding two layers: the "core layer" and the "chosen layer", as depicted in Figure 11.4. This interconnection emphasizes the importance of aligning architecture perspective and organizational readiness to achieve sustainable implementation of Industry 4.0 technologies.

REFERENCES

Bai, C., Dallasega, P., Orzes, G., and Sarkis, J., (2020), "Industry 4.0 Technologies Assessment: a Sustainability Perspective", *International Journal of Production Economics*, Vol. 229, 107776, 2020.

Frank, A. G., Dalenogare, L. S., and Ayala, N. F., (2019), "Industry 4.0 Technologies: Implementation Patterns in Manufacturing Companies", *International Journal of Production Economics*, Vol. 210, pp. 15–26.

Garbie, I., and Garbie, A., (2020a), "Outlook of Requirements of Manufacturing Systems for 4.0", *3rd International Conference of Advances in Science and Engineering Technology (Multi-Conferences) ASET 2020*, February 4–6, 2020, Dubai, UAE, 5 pp.

Garbie, I., and Garbie, A., (2020b), "A New Analysis and Investigation of Sustainable Manufacturing through a Perspective Approach", *3rd International Conference of Advances in Science and Engineering Technology (Multi-Conferences) ASET 2020*, February 4–6, 2020, Dubai, UAE, 5 pp.

Garbie, I., and Garbie, A., (2020c), "Sustainability and Manufacturing: A Conceptual Approach", *Proceedings of the Industrial and Systems Engineering Research Conference (IISE Annual Conference and Expo 2020)*, November 1–3, 2020, New Orleans, LA, USA, 6 pp.

Garbie, I. H., and Garbie, A., (2022), "Identifying and Estimating the Implementation Level for Industry 4.0", *4th International Conference of Advances in Science and Engineering Technology (Multi-Conferences) ASET 2022*, February 21–24, 2022, Dubai, UAE, 6 pp.

Garbie, I., and Garbie, A., (2024a), "Toward Smart Manufacturing Systems Incorporating Reconfiguration Issues", *International Journal of Industrial and Systems Engineering*, Vol. 46, No. 1, pp. 1–33.

Garbie, I., and Garbie, M., (2024b), "Sustainability of Manufacturing Enterprises: A s Psychological Perspective of Linking Manufacturing Layers", *Proceedings of the Industrial and Systems Engineering Research Conference (IISE Annual Conference and Expo 2024)*, May 18–21, 2024, Montreal, Canada, 6 pp.

Garbie, I., and Parsaei, H., (2022), *Reconfigurable Manufacturing Enterprises for Industry 4.0*. CRC Press-Taylor and Francis, 1st Edition. Boca Raton, FL, USA.

Garbie, I. H., Parsaei, H., and Garbie, M., (2023), "Site Readiness Index for Implementing Industry 4.0 based on Design Principles and Emerging Manufacturing Technologies", *Proceedings of the Industrial and Systems Engineering Research Conference (IISE Annual Conference and Expo 2023)*, May 20–23, 2023, New Orleans, Louisiana, USA, 6 pp.

Moeuf, A., Lamouri, R., Pellerin, R., Giraldo, S. T., Valencia, E. T., and Eburdy, R., (2019), "Identification of Critical Success Factors, Risks and Opportunities of Industry 4.0 in SMEs", *International Journal of Production Research*, Vol. 58, no. 5, pp. 1384–1400.

Moktadir, M. A., Alib, S. M., Kusi-Sarpongc, S., and Shaikh, M. A., (2018), "Assessing Challenges for Implementing Industry 4.0: Implications for Process Safety and Environmental Protection", *Process Safety and Environmental Protection*, Vol. 117, pp. 730–741.

Sivananda, D. K., Paranitharan, K. P., and Agniveesh, I. A. (2020), "Interpretive Framework by Analyzing the Enablers for Implementation of Industry 4.0: an ISM Approach", *Total Quality Management and Business Excellence*, doi:10.1080/14783363.2020.1735933

Wang, A., Wan, J. Li, D, and Zhang, C., (2016), "Implementing Smart Factory of Industrie 4.0: An Outlook", *International Journal of Distributed Sensor Networks*, Vol. 2016, 3159805.

12 Industry 4.0 Sustainability

12.1 INTRODUCTION

It can be shown in Figure 12.1 that there are three different layers of the architecture perspective of Industry 4.0/smart manufacturing (Garbie and Garbie, 2024a). The primary layer is so-called the core and is dependable for available emerging advanced manufacturing technologies for I4.0. The second layer is the chosen layer, represented by the manufacturing technologies I4.0 selection with which feasible and convenient manufacturing technologies will be adopted. The final layer could be a given one and is utilized to demonstrate which site readiness can be given based on the integration of physical and virtual systems (Garbie and Garbie, 2024b). This chapter aims to discover the sustainability index for these streams individually first and the architecture perspective interface between them. In this chapter, the manufacturing technologies of these three layers will be distinguished and analyzed, and the sustainability index linking these layers through the architectural perspective will be proposed and outlined independently and universally. All of these issues were discussed from different technical perspectives toward reconfiguration (Garbie and Garbie, 2024).

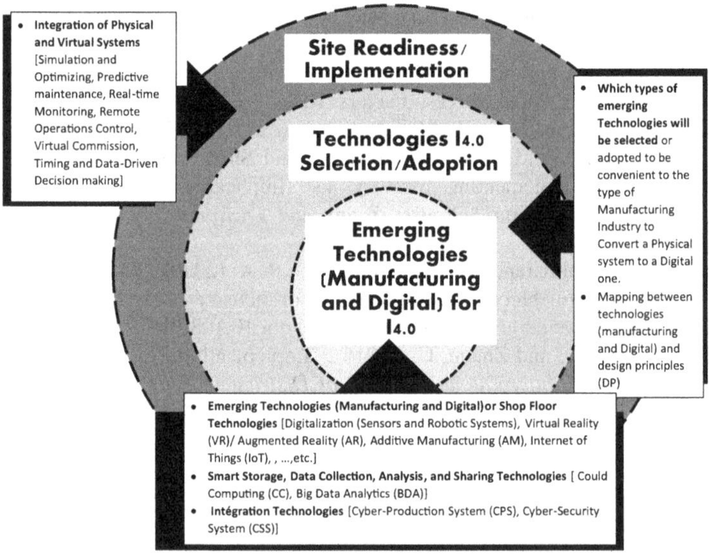

FIGURE 12.1 Layers of architecture perspective in Industry 4.0.

DOI: 10.1201/9781003644835-16

12.2 INDUSTRY 4.0/SMART MANUFACTURING STREAMS

The primary goal of this chapter is to analyze and explore Industry 4.0 (I4.0) and smart manufacturing by identifying the key enablers within each stream. Numerical models are proposed to evaluate indexes related to emerging manufacturing technologies, technology selection for I4.0, and site readiness. Each stream is characterized and distinguished based on these enabling factors (Figure 12.1).

12.2.1 Emerging Technologies (ET) Sustainability

As illustrated in Figure 12.1, the emerging technologies (ET) stream of Industry 4.0 and smart manufacturing can be categorized into three fundamental sub-streams:

1. Emerging manufacturing technologies (shop floor technologies)
2. Smart storage, data collection, analysis, and sharing technologies
3. Integration technologies

Each sub-stream encompasses unique technological characteristics. The I4.0 technologies, as described in Chapter 9, include but are not limited to:

- digitalization (e.g., information and communication technology, Sensors, and Robotic Systems) (f_1),
- industrial Internet of things (f_2),
- big data analytics (f_3),
- cloud computing/cloud manufacturing (f_4),
- cyber-security system (f_5),
- virtual reality/augmented reality (f_6),
- additive manufacturing (f_7),
- cyber-production physical system (f_8), … n, etc.

The sustainability level of the Industry 4.0/smart manufacturing concerns at any given time t for the core stream "emerging technologies for I4.0", $S_{\mathrm{ET}}(t)$, can be expressed as a function of all the aforementioned technologies within the following Equation (9.1) reiterated here as Equation (12.1). This equation is the foundational expression for sustainable manufacturing in the core stream.

$$S_{\mathrm{ET}}(t) = f(f_1, f_2, f_3, \ldots f_n) = \begin{Bmatrix} f_1 \\ f_2 \\ f_3 \\ \text{—} \\ f_n \end{Bmatrix} = \sum_{i=1}^{n} f_i(t)/n \tag{12.1}$$

where:

$f_i(t)$ in Equation (12.1) can be calculated using the following Equation (12.2)

$$f_i(t) = \frac{f_i(t)\text{current} - f_i(t)\min}{f_i(t)\max - f_i(t)\min} \tag{12.2}$$

where:

$f_i(t)$ is the normalized value of technology specified with the emerging technology (ET) at any time t,

$f_i(t)$current is the existing value of technology specified with the emerging technology (ET) at any time t,

$f_i(t)$max and $f_i(t)$min are the maximum and minimum values of technology specified with the emerging technology (ET) at any time t, respectively, and

n is the number of technologies in the emerging technology (ET).

12.2.2 Technologies Selection (TS) Sustainability

According to Figure 12.1 and the previous analysis in Chapter 10, the technologies I4.0 selection (TS) is based on the previous layer "core one" which is mainly dependent on the types of emerging technologies. The sustainability of TS depends on design principles (DP) and their associated emerging technologies (ET). These design principles are:

- Interoperability (g_1)
- Virtualization (g_2)
- Decentralization (g_3)
- Real-time capability (g_4)
- Service orientation (g_5)
- Modularity (g_6)

At that point, the sustainability of the Industry 4.0/smart manufacturing concerns technologies selection (TS) at any time t, $S_{\mathrm{TS}}(t)$, it is numerically defined as a function as presented in Equation (10.7) and repeated in Equation (12.3).

$$S_{\mathrm{TS}}(t) = f(g_1, g_2, g_3, \ldots g_6) = \left\{ \begin{array}{c} g_1 \\ g_2 \\ g_3 \\ \text{—} \\ g_6 \end{array} \right\} = \sum_{i=1}^{6} W_i\, g_i(t) \tag{12.3}$$

Equation (12.3) can be quantitatively rewritten in new forms to assess the sustainability index for emerging technologies selection for Industry 4.0 as in Equation (12.4)

$$S_{\mathrm{TS}}(t) = W_{g1} g_1(t) + W_{g2} g_2(t) + W_{g3} g_3(t) + W_{g4} g_4(t) + W_{g5} g_5(t) + W_{g6} g_6(t) \tag{12.4}$$

where:

$S_{\mathrm{TS}}(t)$ = sustainability index of emerging technologies selection for Industry 4.0 based on mapping between emerging technologies and design principles at any time t,

$g_1(t)$ = percentage (%) of interoperability at a given time t,

$g_2(t)$ = percentage (%) of virtualization at a given time t,

$g_3(t)$ = percentage (%) of decentralization at a given time t,

$g_4(t)$ = percentage (%) of using real-time capability at a given time t,
$g_5(t)$ = percentage (%) of using service orientation at a given time t,
$g_6(t)$ = percentage (%) of modularity at a given time t.

$g_1(t)$, $g_2(t)$, $g_3(t)$, $g_4(t)$, $g_5(t)$, and $g_6(t)$ are estimated as shown in the previous Chapter 10. The symbols W_{g1}, W_{g2}, W_{g3}, W_{g4}, W_{g5} and W_{g6} are the relative weights of interoperability, virtualization, decentralization, real-time capability, service orientation, and modularity, respectively.

12.2.3 Site Readiness (SR) Sustainability

The sustainability of the given layer for Industry 4.0 (I4.0) and smart manufacturing follows a similar procedural approach as the previous layers. This stream is referred to as the sustainability of site readiness (SR). It serves as the subsequent outcome of the core layer (emerging technology, ET) and the chosen layer (technology selection, TS). The sustainability of site readiness integrates the outcomes of the prior layers to ensure the seamless implementation and operational readiness of physical and virtual systems within the I4.0 framework.

The sustainability level of Industry 4.0/Smart Manufacturing in terms of site readiness (*SR*) at any time t, $S_{SR}(t)$, can be mathematically expressed based on the type of issues of site readiness as presented in Equation (11.6) and repeated in Equation (12.5):

- technical challenges of implementation (h_1),
- risks (h_2),
- critical success factors (h_3),
- opportunities (h_4), and
- enablers of implementation (h_5)

$$S_{SR}(t) = f(h_1, h_2, h_3, h_4, h_5) = \left\{ \begin{array}{c} h_1 \\ h_2 \\ h_3 \\ h_4 \\ h_5 \end{array} \right\} = \frac{h_1 + h_2 + h_3 + h_4 + h_5}{5} \tag{12.5}$$

The h_1, h_1, h_1, h_1, and h_1 are estimated as shown in the previous Chapter 11.

12.3 SUSTAINABILITY ASSESSMENT OF INDUSTRY 4.0

The overall sustainability index for Industry 4.0 (I4.0) and smart manufacturing is derived from the sustainability indexes of each layer. These streams are core (emerging technologies, ET), chosen (technology selection, TS), and given (site readiness, SR) represented by $S_{ET}(t)$, $S_{TS}(t)$, and $S_{SR}(t)$, respectively to serve as the foundational components.

The sustainability index is calculated by integrating the individual contributions of these streams, thereby providing an assessment of the sustainability level of the entire Industry 4.0 sustainability framework: $S_{\text{ET}}(t)$, $S_{\text{TS}}(t)$, and $S_{\text{SR}}(t)$. At this stage, the sustainability index of Industry 4.0 is mathematically expressed at a given time t, $S_{\text{SI4.0}}(t)$, as the following Equation (12.6) (Garbie and Garbie, 2024a).

$$S_{I4.0}(t) = \left[S_{\text{ET}}(t)\right]^{\frac{1}{\sum_{j=1}^{2} w_j(t) S_j(t)}} = \left[S_{\text{ET}}(t)\right]^{\frac{1}{[w_{\text{TS}}(t) S_{\text{TS}}(t) + w_{\text{SR}}(t) S_{\text{SR}}(t)]}} \tag{12.6}$$

Equation (12.6) is based on the core layer of Industry 4.0 "sustainability of emerging technologies, $S_{\text{ET}}(t)$" which is expressed as a constant value and later sustainability in technologies I4.0 $S_{\text{TS}}(t)$ and site readiness ($S_{\text{SR}}(t)$) as a variable. In this way, Equation (12.6) is considered an exponential function affected by the three layers of achieving sustainability of Industry 4.0/smart manufacturing. Therefore, the sustainability function of Industry 4.0/smart manufacturing at any time t, $S_{\text{SI4.0}}(t)$, has the form which includes $S_{\text{ET}}(t) > 0$ mathematically and practically. The symbols $W_{\text{TS}}(t)$, $W_{\text{SR}}(t)$ are the relative weights for the technologies I4.0 selection and site readiness for adopting Industry 4.0/smart manufacturing at any time t, respectively.

12.4 NUMERICAL EXAMPLE 12.1

This numerical illustration is utilized to demonstrate the proposed methodology for assessing the sustainability index of Industry 4.0/smart manufacturing concerning the architecture perspective based on emerging technologies (ET), technologies selection (TS), and site readiness (SR). The distinguishing indicators of sustainability issues and performance measurements (existing values and target values (minimum and maximum)) are outlined in Tables 12.1 and 12.2 for the emerging technologies and technologies selection, separately.

12.4.1 For the Emerging Technologies (ET)

Data for the emerging technologies (ET) listed in Table 12.1 are used to normalize and calculate the sustainability index for the emerging technologies and later it helps to identify the mapping with the design principles in the technologies selection (TS). The sustainability index for ET appears in Table 12.2.

12.4.2 For the Technologies Selection (TS)

Data for emerging technologies is the same as that given in Table 12.1 and the associated result is illustrated in Table 12.2. For mapping the design principles and emerging technologies for Industry 4.0, Equations (10.1)–(10.6), for interoperability (g_1), virtualization (g_2), decentralization (g_3), real-time capability (g_4), service orientation (g_5), and modularity (g_6), respectively, as follows:

TABLE 12.1
Data for the Emerging Technologies

Indicator (f)	Units Performance Measures (%)	Performance Metrics Existing Value	Target Value Max	Min
f_1	Percentage (%) of using Digitalization (DI) at a given time t	85	90	80
f_2	Percentage (%) of using the (IIoT) at a given time t	95	100	90
f_3	Percentage (%) of using BDA) at a given time t	85	90	80
f_4	Percentage (%) of using (CC/CM) at a given time t	85	95	80
f_5	Percentage (%) of using (CSS) at a given time t	95	100	90
f_6	Percentage (%) of using (VR/AR) at a given time t	80	90	75
f_7	Percentage (%) of using (AM) at a given time t	95	100	80
f_8	Percentage (%) of using cyber-physical production system (CPPS) at a given time t	95	100	80

TABLE 12.2
Sustainability Index in Emerging Manufacturing Technologies

Indicator (f)	f_1	f_2	f_3	f_4	f_5	f_6	f_7	f_8
Normalized index	0.50	0.50	0.33	0.25	0.50	0.33	0.75	0.75
$S_{ET}(t)$	0.4887							

$$g_1 = \frac{f_1 + f_2 + f_3 + f_4 + f_5 + f_7 + f_8}{7}$$

$$= \frac{0.50 + 0.50 + 0.33 + 0.25 + 0.50 + 0.75 + 0.75}{7} = 0.51$$

$$g_2 = \frac{f_1 + f_5}{2} = \frac{0.50 + 0.50}{2} = 0.50$$

$$g_3 = \frac{f_1 + f_7 + f_8}{3} = \frac{0.50 + 0.75 + 0.75}{3} = 0.67$$

$$g_4 = \frac{f_1 + f_2 + f_3 + f_6}{4} = \frac{0.50 + 0.50 + 0.33 + 0.33}{4} = 0.42$$

$$g_5 = \frac{f_1 + f_2 + f_3 + f_4 + f_6}{5} = \frac{0.50 + 0.50 + 0.33 + 0.25 + 0.33}{5} = 0.39$$

$$g_6 = \frac{f_1 + f_4 + f_5 + f_7}{4} = \frac{0.50 + 0.25 + 0.50 + 0.75}{4} = 0.50$$

Equation (10.8) in Chapter 10 is used to assess the sustainability index for emerging technologies selection for Industry 4.0 as follows, estimating the relative weights between interoperability, virtualization, decentralization, real-time capability, service orientation, and modularity by using the Analytical Hierarchy Processes (AHP). The relatives were estimated as shown in Figure 10.5 as 21.45%, 17.83%, 16.05%, 20.34%, 11.83%, and 12.50% for interoperability, virtualization, decentralization, real-time capability, service orientation, and modularity, respectively. The sustainability index of technologies selection (TS) at any time t, $s_{TS}(t)$ is evaluated by utilizing Equation (10.8), and the results are 0.50 (50.00%).

$$S_{TS}(t) = 0.2145g_1(t) + 0.1783g_2(t) + 0.1605g_3(t) + 0.2034g_4(t)$$
$$+0.1183g_5(t) + 0.1250g_6(t)$$

$$S_{TS}(t) = 0.2145(0.51) + 0.1783(0.50) + 0.1605(0.67) + 0.2034(0.42)$$
$$+0.1183(0.39) + 0.1250(0.50) = 0.50$$

12.4.3 For the Site Readiness (SR)

The illustrated example presented in Section 11.4 (Chapter 11) is used to demonstrate the proposed framework to assess the sustainability index toward I4.0 with respect to site readiness (SR). Data related to technical challenges (h_1), risks (h_2), critical success factors (h_3), using of opportunities (h_4), and enablers of implementation (h_5) are presented and illustrated in Tables 11.6–11.10, respectively. The existing current value of $h_1\left(E^{h_1}\right)$ for technical challenges; the existing current value of $h_2(E^{h_2})$; the existing current value of $h_3(E^{h_3})$; the existing current value of $h_4(E^{h_4})$; and the existing current value of $h_5\left(E^{h_5}\right)$ are estimated using Equations (11.1)–(11.5) and data in Tables 11.6–11.10 as follows.

$$E^{h_1} = \frac{3\times 3 + 4\times 7 + 5\times 4}{70} = 0.8142$$

$$E^{h_2} = \frac{1\times 2 + 2\times 3 + 3\times 4 + 1\times 5}{50} = 0.60$$

$$E^{h_3} = \frac{2\times 3 + 2\times 4 + 4\times 5}{40} = 0.85$$

$$E^{h_4} = \frac{2\times 4 + 3\times 5}{25} = 0.92$$

$$E^{h_5} = \frac{2\times2+4\times6+5\times6}{70} = 0.8285$$

The total site readiness issues result appears in Table 12.3. The sustainability index of site readiness (*SR*) at any time *t*, $s_{SR}(t)$ is evaluated by utilizing Equation (12.5), and the results are outlined in Table 12.3 as 80.25%.

The index of sustainability of Industry 4.0/smart manufacturing is assessed, as specified in Equation (12.6) based on the sustainability of emerging manufacturing technologies. Technologies selection and site readiness have equal relative weights as follows. This implies that the sustainability index in Industry 4.0 to 33.3%

$$S_{I4.0}(t) = [0.4887]^{\frac{1}{[0.50x0.50+0.50x0.8025]}} = 0.333$$

It can be observed from the results in Table 12.4 that when the index of emerging technologies (ET) is adjusted independently at constant intervals while maintaining the values for technologies selection (TS) and site readiness (SR) constant, the overall sustainability index exhibits corresponding changes. This demonstrates the reliance of the sustainability index on variations within the ET layer. Furthermore, as shown in Table 12.5, the highest sustainability index value among the three streams is identified, providing insights into the most influential factor contributing to the overall sustainability of Industry 4.0 and smart manufacturing.

TABLE 12.3
Sustainability Index in Site Readiness

Indicator (h)	h_1	h_2	h_3	h_4	h_5
Index of existing value	81.42	60.00	85.00	92.00	82.85
$S_{SR}(t)$	80.25				

TABLE 12.4
Effects of the Values of Individual Sustainability on the Whole Sustainability

Iteration ($)	$S_{ET}(t)$	$w_{TS}(t)$	$S_{TS}(t)$	$w_{SR}(t)$	$S_{SR}(t)$	$S_{I4.0}$ index
1	0.4887	0.50	0.50	0.50	0.8025	0.3330
2	0.5887	0.50	0.50	0.50	0.8025	0.4433
3	0.6887	0.50	0.50	0.50	0.8025	0.5640
4	0.7887	0.50	0.50	0.50	0.8025	0.6945
5	0.8887	0.50	0.50	0.50	0.8025	0.8343
6	0.9887	0.50	0.50	0.50	0.8025	0.9830

TABLE 12.5
Effects of the Values of Individual Sustainability on the Whole Sustainable Manufacturing

Iteration	$S_{ET}(t)$	$w_{TS}(t)$	$S_{TS}(t)$	$w_{SR}(t)$	$S_{SR}(t)$	$S_{I4.0}(t)$ Index	$S_{I4.0}(t)$ 100%
1	0.9887	0.50	0.50	0.50	0.8025	0.9830	98.30

12.5 CONCLUDING REMARKS

In this chapter, the architecture perspective of connecting Industry 4.0/smart manufacturing layers was proposed and analyzed through the three primary layers of sustainability. The concepts of the core, chosen, and given layers were based on several unique characteristics of elements such as novel and inventive approaches. In each layer, the essential elements were recognized and identified, and the sustainability index was assessed separately. Afterward, the global sustainability index of Industry 4.0/smart manufacturing was also verified and assessed through the numerical scientific model to depict the impact of each layer on the others. It was noticed, logically, that there's a mental interface between the layers of Industry 4.0/smart manufacturing even though this is often a solid world or so-called Industry world.

The architecture system plays a crucial role in Industry 4.0/ smart manufacturing by influencing not only the behavior and decision-making of individuals within the manufacturing enterprise but also in layers themselves. Some effects of the architecture system in Industry 4.0/smart manufacturing will be recommended to be investigated and assessed in real-life case studies such as employee motivation; decision-making; innovation and creativity; behavior change; and organizational culture.

REFERENCES

Garbie, I., and Garbie, A., (2024), "Toward Smart Manufacturing Systems incorporating Reconfiguration Issues", *International Journal of Industrial and Systems Engineering*, Vol. 46, No. 1, pp. 1–33.

Garbie, I., and Garbie, M., (2024a), "Sustainability of Manufacturing Enterprises: As Psychological Perspective of Linking Manufacturing Layers", *Proceedings of the Industrial and Systems Engineering Research Conference (IISE Annual Conference and Expo 2024)*, May 18–21, 2024, Montreal, Canada, 6 pp.

Garbie, I., and Garbie, M., (2024b), "Designing Manufacturing Enterprises System for Industry 4.0", *Proceedings of the Industrial and Systems Engineering Research Conference (IISE Annual Conference and Expo 2024)*, May 18–21, 2024, Montreal, Canada, 6 pp.

Part 5

Sustainability 4.0

13 Effects of Industry 4.0 Technologies on Sustainability

13.1 INTRODUCTION

Talking about sustainability/sustainable development in the manufacturing industry today is unlike talking from 10 or 20 years ago because of Industry 4.0 (I4.0). Therefore, sustainability, nowadays, is called sustainability 4.0 (S 4.0) and is different from previous concepts because of the comprehensive usage of emerging technologies (advanced manufacturing and digital technologies) especially the smart ones raised with the period of Industry 4.0. Understanding the interconnections between Industry 4.0 technologies and sustainability is a complex but essential task.

In the manufacturing domain, Industry 4.0 technologies influence sustainability pillars just as much as these pillars shape the implementation of Industry 4.0. Therefore, the question of precedence is "whether Industry 4.0 or sustainability comes first". The answer will be confusing. For this reason, it is recommended to say "Both". In this chapter, it is important to investigate the effect of Industry 4.0 technologies on the sustainability/sustainable development (SM4.0) index through their sustainability pillars and enablers, the so-called "S4.0". Industry 4.0 represents the fourth industrial revolution, introducing transformative technologies such as advanced process technology (Lins and Oliveira, 2020).

Emerging Industry 4.0 technologies—such as the Industrial Internet of Things (IoT), blockchain, cloud computing/manufacturing (CC/CM), additive manufacturing (AM), artificial intelligence (AI), big data analytics (BDA), and simulation and augmented/virtual reality (AR/VR)—enhance sustainability in manufacturing by fostering innovation, flexibility, efficiency, and productivity. Furthermore, communication, information, and intelligence technologies bolster sustainability by addressing global competition, volatile market demands, and increasing personalization (Bai et al., 2020). This represents the economic sustainability of adopting Industry 4.0 technologies.

Industry 4.0 technologies contribute to social sustainability by enhancing employee satisfaction and motivation, often through ergonomic systems tailored to individual needs. However, challenges such as unemployment, data security, electronic waste, and complexity must be addressed. On the environmental front, Industry 4.0 technologies reduce energy and resource consumption via data analytics and intelligent systems, ultimately minimizing carbon emissions. These technologies also mitigate the ecological imbalance and environmental degradation associated with traditional manufacturing processes, including global warming and pollution.

DOI: 10.1201/9781003644835-18

13.2 SUSTAINABILITY PILLARS OF MANUFACTURING INDUSTRIES

Sustainability in manufacturing industries is founded on the triple bottom line (TBL) framework, integrating economic, social, and environmental dimensions (Figure 13.1). TBL serves as a comprehensive performance measurement system for manufacturing industries, and its integration with Industry 4.0 technologies has gained prominence in recent years (Margherita and Braccini, 2018).

13.2.1 Pillar Integration and Features

The integration of sustainability pillars generates unique features categorized as short-term, long-term, and continuous:

13.2.2 Short-Term Features

- **Socio-Economic**: Job creation, skills enhancement, local economic impacts, social investments, business ethics, and tax/royalty contributions.
- **Social-Environmental**: Safety and health, global climate change mitigation, local environmental impacts, and resource management.
- **Eco-Efficiency**: Life-cycle analysis, product stewardship, and resource efficiency.

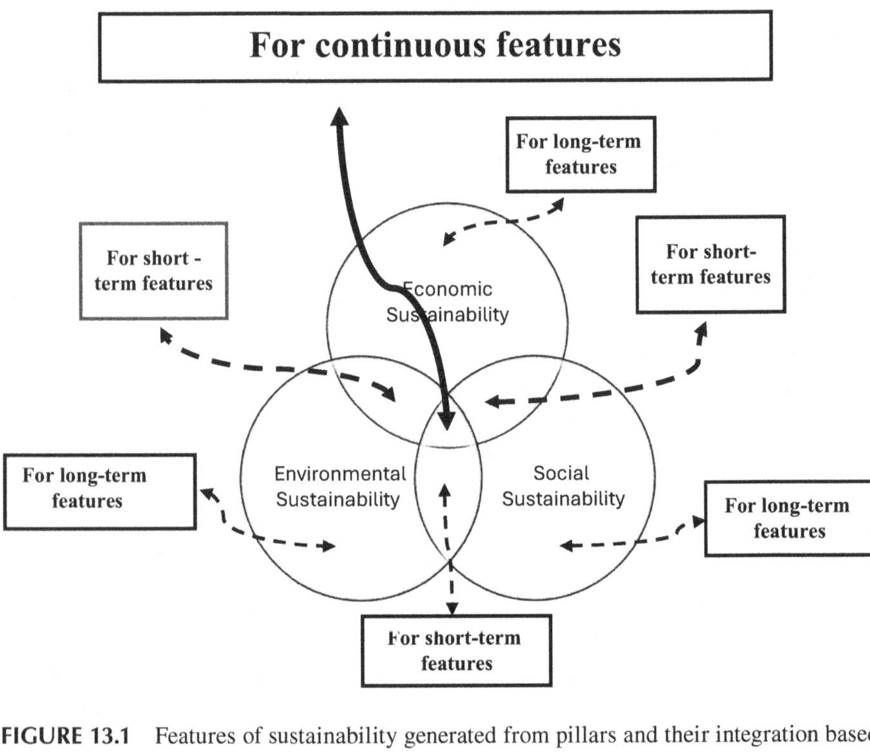

FIGURE 13.1 Features of sustainability generated from pillars and their integration based on the TBL.

13.2.3 Long-Term Features

- **Economics**: Revenue growth, earnings, net cash flow, and shareholder returns.
- **Social**: Labor standards, employment diversity, and community dialogue.
- **Environmental**: Waste minimization, emissions reduction, ecosystem services, biodiversity preservation, and regulatory compliance.
- **Continuous Features**: Integrating all three pillars fosters value creation, enhancing well-being and welfare.

13.2.4 Pillar Issues, Indicators, and Metrics

Each sustainability pillar encompasses specific issues/aspects with associated indicators and performance metrics (Tables 13.1–13.3). Integrating these pillars is a recognized priority in manufacturing research, emphasizing sustainability as an engineering responsibility (Garbie, 2013, 2014, 2015).

TABLE 13.1
Economic Pillar and its Associated KPIs and Performance Units

Issue/Aspect	Key Performance Indicator (KPI)	Performance Measures (Units)
Globalization issues	Logistics and supply chain	Number of disturbances by suppliers
	Information technology	Possibility (%) of using the high Internet speed
	Power price	$/Gallen
	Contemporary markets	New markets
	Commercial business models	Generating customers
Contemporary issues	Operational technology	Possibility (%) of updating existing technology
	Compliance with regulations	Possibility (%) of following the government regulations
	Populations growth	Percentage (%) of increasing populations
	Economics enhancement	Percentage (%) of profitability
	Natural resources depletion	Possibility (%) of minimizing resource depletion
Innovation	How to satisfy needs	Possibility (%) to estimate the needs
	Exploit the opportunity of market	Possibility (%) to exploit the market opportunity
	Cost of innovative/renovative product	$/item
	Time to innovative/renovative product	Unit time/item
	Development flexibility	Possibility (%) of adapting resources inside a plant
	Customized products for certain areas	Number of areas around the world
	Customized products for specified groups	Number of special groups around the world

(Continued)

TABLE 13.1 (*Continued*)
Economic Pillar and its Associated KPIs and Performance Units

Issue/Aspect	Key Performance Indicator (KPI)	Performance Measures (Units)
Design and Reconfiguration	Facilities design size	Number of machines and equipment
	Facilities design flexibility	Number of distinct manufacturing processes
	Types of logistics systems inside the facility	Number of distinct logistics systems resources
	Warehouse storage system	Cubic feet
	Identification and data capture	Number of data capture system
	Facilities location	Number of facilities around the world
	Functional or processes layout	Number of reminder and functional cells
	Product layout	Number of dedicated cells
	Focused layout	Number of assigned and grouped cellular system
Competitive policies	Optimizing complexity	Possibility (%) of optimizing complexity
	Maximizing leanness	Possibility (%) of increasing the value added
	Maximizing agility	Possibility (%) of increasing manufacturing agility
	Maximizing remanufacturing	Possibility (%) of increasing the number of components that can be replaced again
	Maximizing recycling	Possibility (%) of increasing the number of components that can be recycled again
Performance analyses	Manufacturing cost	$/product
	Manufacturing lead time	Unit time (Month, week, day, hour)
	Production rate	Products/hour
	Employees performance	Productivity of employee
	Reliability and maintainability	Availability of machines (%)
	Quality control	Percentage (%) of rejected products
Flexible organization management	Planning	Possibility (%) of compliance strategic planning
	Organizing work	Possibility (%) of subordinates per supervisor/superintended
	Structure of organization	Number of organization structure
	Leadership role	Degree (%) of leadership
	Staffing	Percentage (%) to access to professional employees

13.3 INDUSTRY 4.0/SMART MANUFACTURING

13.3.1 Industry 4.0 Technologies Overview

Industry 4.0 represents a suite of transformative technologies that have emerged since the beginning of the 21st century. The primary goal of adopting Industry 4.0 technologies is to integrate advanced digital and manufacturing technologies

TABLE 13.2
Social Pillar and its Associated KPIs and Performance Units

Issue or Aspect	Key Performance Indicators (KPIs)	Performance Measures (units)
Workplace management	Employment	Number of hired new employees
	Workplace conditions	Percentage (%) of suitability of workplace
	Stakeholders' negotiations	Possibility (%) of social dialogue
	Social safety	Percentage (%) of social safety
	Professional development	Percentage (%) of developing employees
Employee rights	Hiring children	Possibility (%) of eliminating hiring children
	Building union	Possibility (%) of installing employees' union
	Diversity and equity	Possibility (%) of eliminating discrimination
Societal commitment	Participation in local Community	Possibility (%) of involvement in the local community
	Education level	Possibility (%) of enhanced education level
	Healthcare level	Possibility (%) of enhanced healthcare level
	Job creation level	Percentage (%) of creating jobs
	Corporate social responsibility (CSR)	Percentage (%) of investment in society
	Culture transfer	Possibility (%) of fostering culture in society
Customers' requirements	Exploring marketing	Percentage (%)
	Protection of private life	Percentage (%)
	Easy to access services	Percentage (%)
Business requirements	Corruption solving	Percentage (%)
	Transparency in business	Percentage (%)
	International and world cultures' understanding	Percentage (%)

with traditional manufacturing technologies in cyber-physical production systems (CPPS). Despite challenges such as high capital costs, unemployment concerns, the need for skilled workers, and gaps in management systems and knowledge, both academicians and practitioners advocate for their adoption due to the potential for value creation and performance enhancement.

However, the high implementation costs often inhibit small- and medium-sized enterprises (SMEs) from adopting these technologies. In addition, many manufacturing industries lack readiness for full implementation but view Industry 4.0 as an opportunity to bypass traditional industrial stages and achieve leapfrogging development.

13.3.2 Adoption beyond Automation

Implementing Industry 4.0 extends beyond industrial automation. It is a combination of all emerging technologies reserved for advanced manufacturing and digital technologies. It involves utilizing a comprehensive range of these advanced technologies, including:

TABLE 13.3
Environmental Pillar and its Associated KPIs and Performance Units

Issue or Aspect	Key Performance Indicators (KPIs)	Performance Measures (units)
Environment management	Budget assigned for Environment	Budget cost for environmental compliance
	Certification	Possibility of getting ISO14001
	Compliance with Environmental issues	Possibility of assessing the environmental impact
	Implications effects on workers	Number of accidents due to environment
Use of resources	Renewable energy	Percentage (%) of using renewable energy
	Recycled water	Percentage (%) of using recycled water
	Recyclable wastes	Percentage (%) of using recycled wastes
Pollution	Pollution of air	Percentage (%) of carbon dioxide in air
	Pollution of water	Percentage (%) of dirty particles in the water
	Pollution of Land	Percentage (%) of landfill location for useful land
Dangerousness	Inputs dangerous materials	Percentage (%) of dangerous materials getting in manufacturing processes
	Output dangerous materials	Percentage (%) of dangerous materials getting out of manufacturing processes
	Dangerous wastes	Percentage (%) of dangerous wastes
Natural environment	Eco-system services	Percentage (%) of carbon dioxide
	Biodiversity	Percentage (%) of the health of ecosystems
	Land use	Land used for the factory/plant
	Development of rural areas	Possibility of investment in rural areas

- **Digital and Manufacturing Technologies**: Internet of Things (IoT), cloud computing/manufacturing (CC/CM), artificial intelligence (AI), virtual/augmented reality (VR/AR), blockchain, additive manufacturing (AM), and robotics.
- **Applications: Supply Chain Management** (Lee et al., 2017; Kuo and Kusiak, 2019; Hammera et al., 2017; Santos et al., 2017), information and communication systems (Ehie and Chilton, 2021; Lins and Oliveira, 2020; Yang et al., 2019; Trappey et al., 2017; Ardanza et al., 2019; Ding et al., 2019), and advanced sensors and actuators in process technology (Lins and Oliveira, 2020; Yang et al., 2019).

Innovation, a critical enabler of sustainability, is supported significantly by Industry 4.0 technologies. These technologies reduce product development time and enable customization of goods and services to regional and personalized manufacturing needs (Lee et al., 2017; Kuo and Kusiak, 2019; Hammera et al., 2017; Santos et al., 2017).

13.3.3 Digitization and Big Data

Digitization and big data analytics play a crucial role in scaling and evaluating enterprise functionality and flexibility (Lee et al., 2017; Kuo and Kusiak, 2019; Hammera et al., 2017; Santos et al., 2017). Key applications include:

- Assigning enterprise size and operational scope (Qin and Cheng, 2017; Alavian et al., 2019; Zhou et al., 2012).
- Enhancing material handling equipment (Ding et al., 2019).
- Optimizing identification systems (Ardanza et al., 2019).

13.3.4 Impact on Sustainability

Investments in Industry 4.0 technologies significantly influence sustainability by merging advanced digital and manufacturing innovations. Industry 4.0 facilitates the development of intelligent systems that integrate lean and agile manufacturing strategies to enhance efficiency and effectiveness.

- **Lean Production and IoT**: The IoT, combined with intelligent technologies such as sensors and actuators, is widely implemented to upgrade legacy machines, enabling data collection and enhancing design efficiency (Huang et al., 2022; Psomas and Deliou, 2024; Rahardjo et al., 2023; Kassem et al., 2024).
- **Complexity Reduction**: Traditional manufacturing complexities are optimized or eliminated through Industry 4.0 technologies (Mourtzis et al., 2019).
- **Illustrating the Effects**: Figures 13.2–13.4 present the effects of Industry 4.0 technologies on the three sustainability pillars:

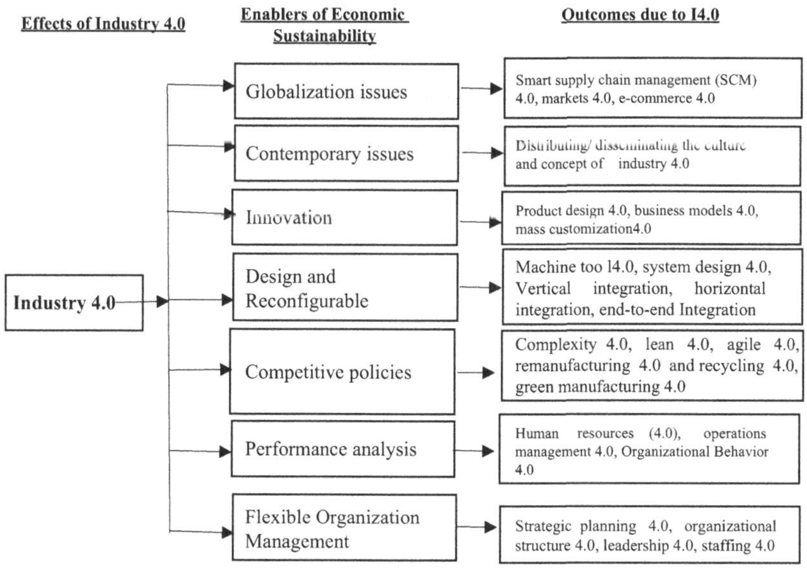

FIGURE 13.2 Effect of Industry 4.0 technologies on enablers of economic sustainability.

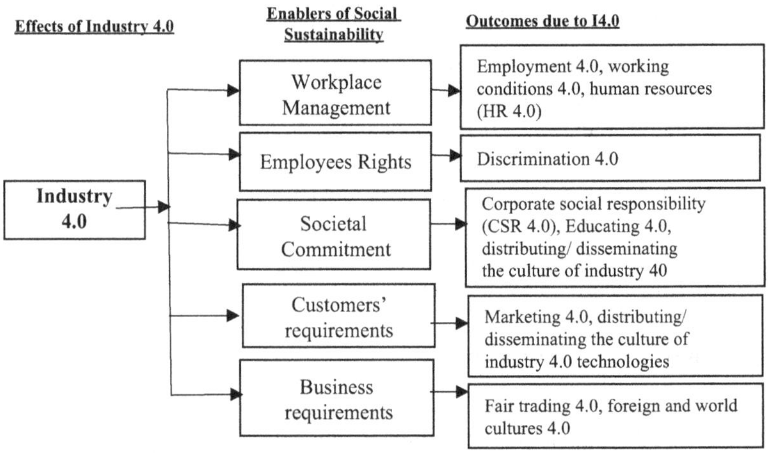

FIGURE 13.3 Effect of Industry 4.0 on enablers of social sustainability.

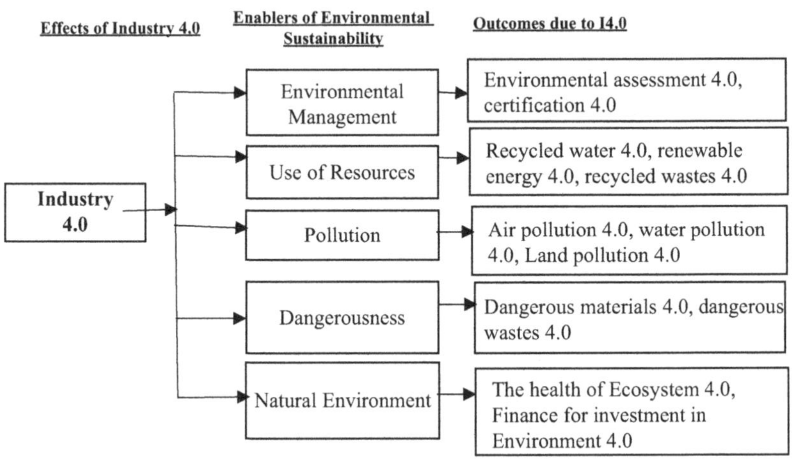

FIGURE 13.4 Effect of Industry 4.0 on enablers of environmental sustainability.

- **Economic Sustainability**: Enhancing cost-effectiveness, productivity, and innovation-driven revenue streams.
- **Social Sustainability**: Improving employee satisfaction, safety, and workforce adaptability while addressing societal challenges.
- **Environmental Sustainability**: Reducing energy consumption, carbon emissions, and waste through intelligent resource management and eco-efficient processes.

13.4 CONCLUDING REMARKS

This chapter highlights the transformative potential of adopting and implementing Industry 4.0 technologies in manufacturing industries, particularly within sustainable industrial systems. These technologies give rise to new concepts and ideologies across the pillars of sustainability: economic, social, and environmental.

13.4.1 Economic Sustainability

The integration of Industry 4.0 technologies introduces innovative approaches that reshape traditional business models and processes. Key emerging concepts include:

- **SCM 4.0**: Smart supply chain management powered by data-driven decision-making.
- **Innovation and Product Design 4.0**: Advanced tools for customized and sustainable product creation.
- **Business Models 4.0**: Digital transformation-driven strategies enabling adaptability and competitiveness.
- **Mass Customization 4.0**: Flexible manufacturing systems tailored to individual customer demands
- **Machine Tools and System Design 4.0**: Intelligent systems designed for efficiency and Scalability
- **Lean and Agile Manufacturing 4.0**: Enhanced production methodologies minimizing waste and maximizing responsiveness.
- **Operations Management 4.0**: Streamlined workflows integrating IoT, AI, and automation.

13.4.2 Social Sustainability

Industry 4.0 technologies also redefine workforce dynamics and societal impacts. Emerging concepts include:

- **Employment 4.0**: Job creation through upskilling and adaptation to advanced technologies.
- **Working Conditions 4.0**: Ergonomic and safe environments supported by smart systems.
- **Human Resources 4.0 (HR4.0)**: Digital HR practices enhancing workforce engagement and productivity.
- **Educating 4.0**: Advanced training and education to prepare employees for technological integration.
- **Fair Trading 4.0**: Promoting ethical practices in global supply chains.
- **Dissemination of Industry 4.0 Culture**: Encouraging widespread adoption and understanding of I4.0 technologies.
- **Foreign Cultures 4.0**: Adapting systems to accommodate diverse cultural contexts and inclusiveness

13.4.3 Environmental Sustainability

The environmental pillar benefits from the adoption of Industry 4.0 through enhanced resource management and reduced ecological footprints. Key concepts include:

- **Environmental Assessment 4.0**: Real-time monitoring of environmental performance metrics.
- **Recycled Water 4.0**: Advanced systems for water reuse and conservation.
- **Renewable Energy 4.0**: Integration of sustainable energy sources into manufacturing processes.
- **Recycled Wastes 4.0**: Improved waste management systems utilizing smart technologies.
- **Pollution 4.0**: Innovations targeting emissions reduction and pollution control.

REFERENCES

Alavian, P., Eun, Y., Meerkov, S. M., and Zhang, L., (2020), "Smart Production Systems: Automating Decision-Making in a Manufacturing Environment", *International Journal of Production Research*, Vol. 58, No. 3, pp. 828–845.

Ardanza, A., Moreno, A., Segura, A., Cruz, M., and Aguinaga, D., (2019), "Sustainable and Flexible Industrial Human-Machine Interfaces to Support Adaptable Applications in the Industry 4.0 Paradigm", *International Journal of Production Research*, Vol. 57, No.12, pp. 4045–4059.

Bai, C., Dallasega, P., Orzes, G., and Sarkis, J., (2020), "Industry 4.0 Technologies Assessment: a Sustainability Perspective", *International Journal of Production Economics*, Vol. 229, 107776.

Ding, K., Chan, F. T. S., Zhang, X., Zhou., and Zhang, F., (2019), "Defining a Digital Twin-based Cyber-Physical Production System for Autonomous Manufacturing in Smart Shop Floors", *International Journal of Production Research*, Vol. 57, No. 20, pp. 6315–6334.

Ehie, I. C., and Chilton, M. A., (2020), "Understanding the Influence of IT/OT Convergence on the Adoption of Internet of Things (IoT) in Manufacturing Organizations: An Empirical Investigation", *Computers in Industry*, Vol. 115, p. 103166.

Garbie, I., (2013), "DFSME: Design for Sustainable Manufacturing Enterprises (An Economic Viewpoint)," *International Journal of Production Research*, Vol. 51, No. 2, pp. 479–503.

Garbie, I., (2014), "An Analytical Technique to Model and Assess Sustainable Development Index in Manufacturing Enterprises," *International Journal of Production Research*, Vol. 52, No. 16, pp. 4876–4915.

Garbie, I., (2016), "Sustainable Development Index on Macro-Level Industrial Estates and Specified Sectors", *International Journal of Industrial and Systems Engineering*, Vol. 24, No. 3, pp. 384–421.

Hammera, M., Somersb, K., Karrea, H., and Ramsauera, C., (2017), "Profit per Hour as a Target Process Control Parameter for Manufacturing Systems Enabled by Big Data Analytics and Industry 4.0 Infrastructure", *Procedia CIRP*, Vol. 63, pp. 715–720.

Huang, Z., Jowers, C., Kent, D., Manshadi, A. D., and Dargusch, M. S., (2022), "The Implementation of Industry 4.0 in Manufacturing: from Lean Manufacturing to Product Design", *The International Journal of Advanced Manufacturing Technology*, Vol. 121, pp. 3351–3367.

Kassem, B., Callupe, M., Rossi, M., Rossini, M., and Staudache, A. P., (2024), "Lean 4.0: a Systematic Literature Review on the Interaction between Lean Production and Industry 4.0 Pillars", *Journal of Manufacturing Technology Management,* Vol. 35 No. 4, pp. 821–847.

Kuo, Y.-H., and Kusiak, A., (2019), "From Data to Big Data in Production Research: the Past and Future Trends", *International Journal of Production Research*, Vol. 57, Nos. 15–16, pp. 4828–4853.

Lee, J. Y., Yoon, J. S., and Kim, B.-H., (2017), "A Big Data Analytics Platform for Smart Factories in Small and Medium-Sized Manufacturing Enterprises: An Empirical Case Study of a Die Casting Factory", *International Journal of Precision Engineering and Manufacturing*, Vol. 18, No. 10, pp. 1353–1361.

Lins, T., and Oliveira, R. A. R., (2020), "Cyber-Physical Production Systems Retrofitting in Context of Industry 4.0", *Computers and Industrial Engineering*, Vol. 139, 106193.

Margherita, E. G., and Braccini1, A. M., (2023), "Industry 4.0 Technologies in Flexible Manufacturing for Sustainable Organizational Value: Reflections from a Multiple Case Study of Italian Manufacturers", *Information Systems Frontiers*, Vol. 25, No.3, pp. 995–1016.

Mourtzis, D., Fotia, S., Boli, N., and Vlachou, E., (2019), "Modelling and Quantification of Industry 4.0 Manufacturing Complexity based on Information Theory: A Robotics Case Study", *International Journal of Production Research*, Vol. 57, No. 22, pp. 6908–6921.

Psomas, E., and Deliou, C., (2024), "Lean Manufacturing Practices and Industry 4.0 Technologies in Food Manufacturing Companies: the Greek Case", *International Journal of Lean Six Sigma*, Vol. 15, No. 4, pp. 763–786.

Qin, S.-F., and Cheng, K., (2017), "Future Digital Design and Manufacturing: Embracing Industry 4.0 and Beyond", *Chinese Journal of Mechanical Engineering*, Vol. 30, pp.1047–1049.

Rahardjo, B., Wang, F.-K., Yeh, R.-H., and Chen, Y.-P., (2023), "Lean Manufacturing in Industry 4.0: a Smart and Sustainable Manufacturing System", *Machines*, 2023, Vol. 11, pp. 72–89.

Santos, M. Y., Oliveira e, J., Andrade, C., Lima, F. V., Costa, E., Costa, C., Martinho, B., and Galvão, J., (2017) "A Big Data System Supporting Bosch Braga Industry 4.0 Strategy", *International Journal of Information Management*, Vol. 37, pp. 750–760.

Trappey, A. J. C, Trappey, C. V., Govindarajan, U. H., Chuang, A. C., and Sun, J. J., (2017), "A Review of Essential Standards and Patent Landscapes for the Internet of Things: A Key Enabler for Industry 4.0", *Advanced Engineering Informatics*, Vol. 33, pp. 208–229.

Yang, H., Kumara, S., Bukkapatnam, S. T. S., and Tsung, F., (2019), "The Internet of Things for Smart Manufacturing: A Review", *IISE Transactions*, Vol. 51, No. 11, pp. 1190–1216.

Zhou, Z., Xie, S., and Chen, D., (2012), *Fundamentals of Digital Manufacturing Science*. Springer, London.

14 Building Sustainability 4.0

14.1 INTRODUCTION

The integration of Industry 4.0 technologies with sustainability, referred to as Sustainable Manufacturing 4.0 (SM 4.0), has become a critical focus area. Researchers and practitioners are increasingly developing strategies and frameworks to align the innovations of Industry 4.0 with the principles of sustainability and sustainable development. Such integration is vital for advancing both theoretical and practical applications.

Effective implementation of these strategies requires the active involvement of three key stakeholders:

1. **Academicians**, through their universities and research centers, contribute cutting-edge research and innovation.
2. **Practitioners**, through industrial organizations and manufacturing industries, bring practical expertise and application capabilities.
3. **Governmental agencies** play a crucial role by providing regulatory support and policy frameworks to facilitate adoption (Reis et al., 2021).

Figure 14.1 illustrates the interaction and collaboration among these groups, each of which has distinct roles and characteristics essential to the success of S 4.0. These efforts not only support the transition to sustainable practices but also foster the development of a holistic framework for sustainability in Industry 4.0.

The relationship between the technologies of adopting Industry 4.0 and sustainability performance remains relatively unexplored, particularly within the context of the circular economy (CE) (Kamble and Gunasekaran, 2019). A deeper conceptual, analytical, and empirical investigation is required to determine effective strategies for adopting and integrating emerging I4.0 technologies into sustainable manufacturing practices aligned with the circular economy framework. Industrial organizations

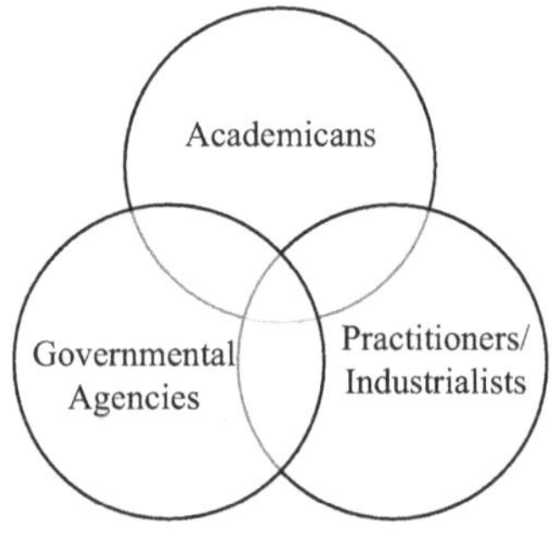

FIGURE 14.1 Integrating academicians, practitioners/industrialists, and government agencies.

 DOI: 10.1201/9781003644835-19

are increasingly recognizing the need to embrace these technologies to enhance their sustainable performance.

It has been observed that developed countries have achieved significantly higher rates of sustainability adoption and implementation in their manufacturing sectors compared to emerging and developing nations (Yadav et al., 2020). In emerging countries, the rate of adoption is modest, while in developing countries, it remains notably weak. Successful implementation of I4.0 technologies is more apparent in large-scale industrial organizations. In contrast, small- and medium-sized enterprises (SMEs), particularly in emerging and developing countries, face considerable challenges in adopting these advanced technologies to achieve sustainability effectively.

14.2 LINKING THE PILLARS OF CORE LAYER OF SUSTAINABILITY 4.0

Academicians (A), practitioners/industrialists (P), and governmental agencies (G) form the foundational pillars of the core layer of Sustainability 4.0 (S 4.0). These pillars play a critical role in bridging the gap between Industry 4.0 technologies and sustainable manufacturing, as illustrated in Figure 14.2. Incorporating these pillars into a mathematical model provides a robust framework to substantiate their interconnectedness. The integration of these pillars in mathematical expression is recommended as solid evidence of this link, which is expressed in Equation (14.1). The integration between these pillars at any given time *t*, $I_{\mathrm{IM4.0}}(t)$, is expressed as a function of A, P, and G and illustrated in Figure 14.2. This will be considered, again, the core layer of Sustainability 4.0 at any given time t [*S*4.0 (*t*)]. The framework, as depicted in Figure 14.2, underscores the dynamic interactions among these key stakeholders, emphasizing their pivotal roles in advancing sustainability goals.

$$I_{\mathrm{IM4.0}}(t) = f(A(t),\ P(t),\ G(t)) = \begin{Bmatrix} A(t) \\ P(t) \\ G(t) \end{Bmatrix} \tag{14.1}$$

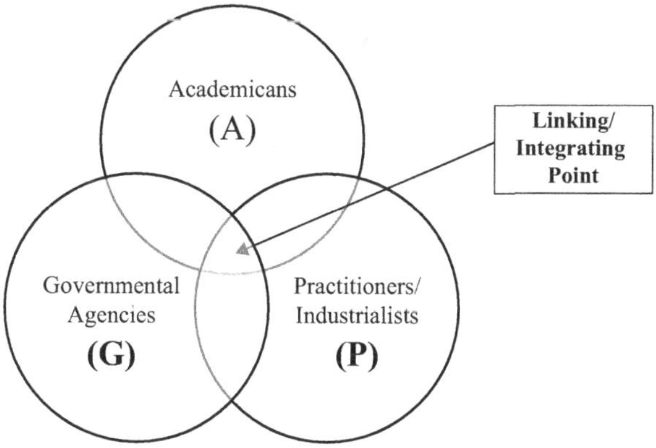

FIGURE 14.2 Linking stakeholders in Industry 4.0 and sustainable manufacturing.

Equation (14.1) is rewritten as Equations (14.2) and (14.3).

$$I_{IM4.0}(t) = \sum_{i=1}^{3} W_i(t) X_i(t) \tag{14.2}$$

$$I_{IM4.0}(t) = W_A(t) X_A(t) + W_P(t) X_P(t) + W_G(t) X_G(t) \tag{14.3}$$

where:

$X_A(t)$ = level of involvement of academicians in linking/integrating at time *t*,

$X_P(t)$ = level of involvement of practitioners/industrialists in linking/integrating at time *t*,

$X_G(t)$ = level of involvement of governmental agencies in linking/integrating at time *t*.

$W_A(t)$, $W_P(t)$, and $W_G(t)$ represent the relative importance of academicians, practitioners/industrialists, and governmental agencies, respectively. These weights are determined using the Analytic Hierarchy Process (AHP), which evaluates the input and experiences of stakeholders or a group of decision-makers.

Stakeholders involved in this process include academicians, senior management, practitioners/industrialists, plant managers, operators, suppliers, and representatives from governmental agencies. AHP provides a systematic approach for deriving these weights by analyzing the relative priorities and contributions of each group, as summarized in the following matrix.

$$A = \begin{bmatrix} 1 & \frac{W_A}{W_P} & \frac{W_A}{W_G} \\ \frac{W_P}{W_A} & 1 & \frac{W_P}{W_G} \\ \frac{W_G}{W_A} & \frac{W_G}{W_P} & 1 \end{bmatrix}$$

14.3 PERSPECTIVE BUILDING OF SUSTAINABILITY 4.0

An architecture perspective, or mental point of view, has been proposed and analyzed to explore the interconnection between manufacturing layers within the context of Sustainability 4.0. This perspective focuses on the three primary streams of sustainable manufacturing (SM), as illustrated in Figure 14.3. The analysis of sustainable manufacturing presented in Chapters 4–8 reveals the existence of an "architecture interface" between the layers of manufacturing. This observation is particularly noteworthy given the traditionally rigid and structured nature of the manufacturing environment, often referred to as the "Industry world".

The architecture perspective system within manufacturing enterprises significantly influences sustainable manufacturing by shaping individual behaviors and decision-making processes. This influence extends to the interactions among the

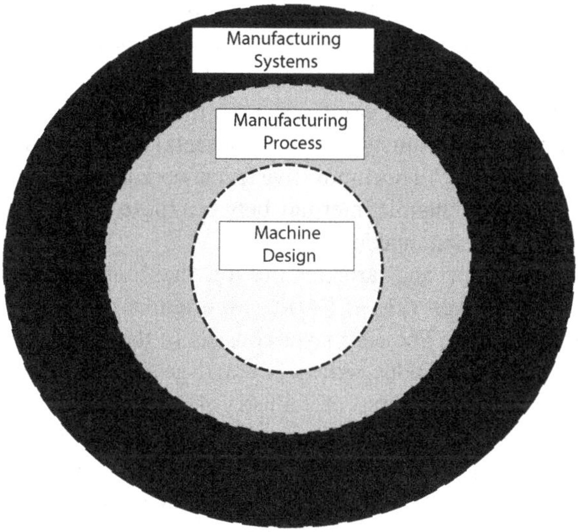

FIGURE 14.3 Sustainable manufacturing layers according to architecture perspective.

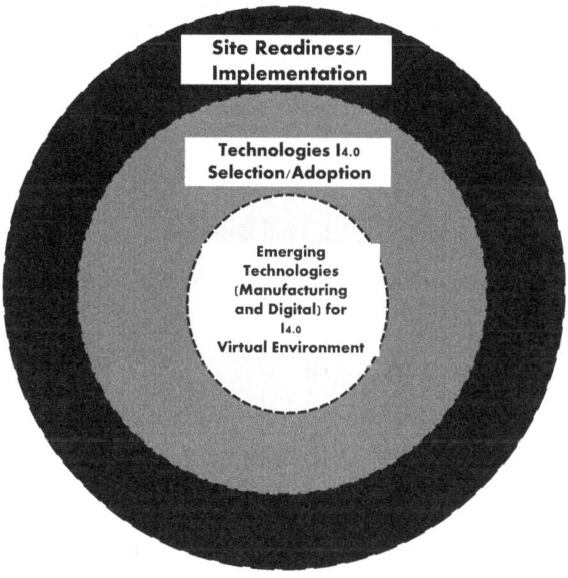

FIGURE 14.4 Industry 4.0 layers according to the architecture perspective.

various manufacturing layers, highlighting the importance of integrating psychological insights into strategies for advancing Sustainability 4.0.

An architecture perspective has also been proposed and analyzed to explore the connection between Industry 4.0 layers and the three primary layers of sustainability, as illustrated in Figure 14.4. The architecture system plays a pivotal role in Industry 4.0 and smart manufacturing, as highlighted in Chapters 9–12. It influences the behavior and decision-making of individuals within manufacturing enterprises and impacts the interactions between various manufacturing layers.

The integration of Industry 4.0 and sustainable manufacturing layers, viewed through an architecture lens and combined with the foundational pillars of academicians, practitioners/industrialists, and governmental agencies, gives rise to a new concept: "Sustainability 4.0". This innovative framework is illustrated in Figures 14.5 and 14.6 demonstrate the dynamic interplay between these elements and their collective potential to advance sustainable practices.

Based on the concept and architecture of the architecture perspective of Sustainability 4.0 at a given time *t*, S4.0(t), Figure 14.6 illustrates the integration layers of Sustainability 4.0. The core layer consists of the foundational pillars: academicians (A), practitioners/industrialists (P), and governmental agencies (G). The chosen layer represents the effects of Industry 4.0 technologies, while the given layer pertains to sustainable manufacturing. This framework addresses the critical question: "Which comes first, or which one affects the other?" The analysis suggests that the effects of Industry 4.0 and smart manufacturing significantly influence sustainability and sustainable development.

To estimate the Sustainability 4.0 at any time t, $S4.0(t)$, including the integration of sustainability of manufacturing, $S_M(t)$ derived from Equation (8.1) and sustainability of Industry 4.0, $S_{I4.0}(t)$, derived from Equation (12.9), is required. The resulting model for $S4.0(t)$ is expressed as a function of $I_{\text{IM4.0}}(t)$, $S_{I4.0}(t)$, and $S_M(t)$ as illustrated in Figure 14.6 and defined by Equation (14.4).

$$S4.0(t) = \left\{ \begin{array}{c} I_{\text{IM4.0}}(t) \\ S_{I4.0}(t) \\ S_M(t) \end{array} \right\} = f\left(I_{\text{IM4.0}}(t),\, S_{I4.0}(t),\, S_M(t)\right) \tag{14.4}$$

Equation (14.4) can be rewritten as Equations (14.5) and (14.6).

$$S4.0(t) = \left[I_{\text{IM4.0}}(t)\right]^{\frac{1}{\left[W_{S_M}(t)\cdot S_M(t) + W_{S_{I4.0}}(t)\cdot S_{I4.0}(t)\right]}} \tag{14.5}$$

$$S4.0(t) = \left[W_A(t)X_A(t) + W_P(t)X_P(t) + W_G(t)X_G(t)\right]^{\frac{1}{\left[W_{S_M}(t)\cdot S_M(t) + W_{S_{I4.0}}(t)\cdot S_{I4.0}(t)\right]}} \tag{14.6}$$

$W_{S_M}(t)$ and $W_{S_{I4.0}}(t)$ are the relative weights of manufacturing sustainability and Industry 4.0 sustainability, respectively.

14.4 NUMERICAL EXAMPLE #1

14.4.1 Linking the Pillars of A, P, and G

The integration of the pillars—academicians (A), practitioners/industrialists (P), and Governmental agencies (G)—is initially estimated as the core layer of Sustainability 4.0 (S4.0). Assuming the levels of involvement for these pillars at time ttt are 90%,

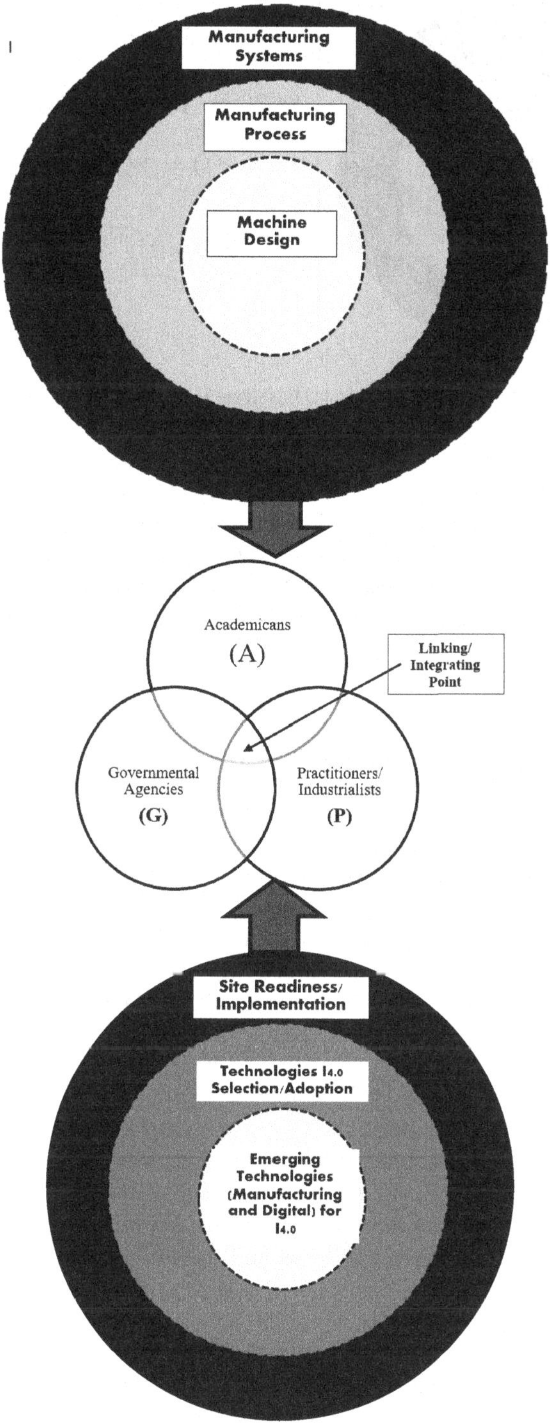

FIGURE 14.5 Integrating Industry 4.0 and sustainable manufacturing with pillars of A, P, and G.

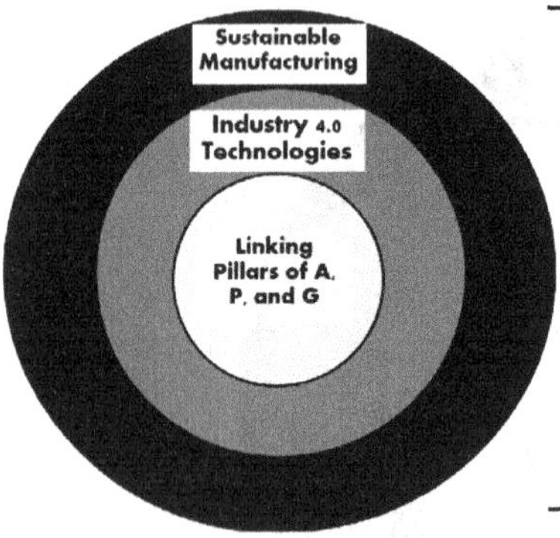

$$S4.0(t) = [I_{IM4.0}(t)]^{\frac{1}{[W_{S_M}(t)\cdot S_M(t)+W_{S_{I4.0}}(t)\cdot S_{I4.0}(t)]}}$$

$$S4.0(t) = [W_A(t)X_A(t) + W_P(t)X_P(t) + W_G(t)X_G(t)]^{\frac{1}{[W_{S_M}(t)\cdot S_M(t)+W_{S_{I4.0}}(t)\cdot S_{I4.0}(t)]}}$$

FIGURE 14.6 Building Sustainability 4.0 based on the architecture perspective.

80%, and 60%, respectively, their integration is calculated using Equation (14.3). For this estimation, equal relative weights are assigned to each of the major pillars. Based on the assumed values for A, P, and G in terms of their degrees or levels of involvement and their equal relative weights, the sustainability of linking these pillars is calculated to be 76.50%. This result is categorized as medium or approaching a high level of integration, indicating a strong foundation for the core layer of S4.0.

$$I_{\text{IM4.0}}(t) = 0.333\cdot 0.90 + 0.333\cdot 0.80 + 0.333\cdot 0.60 = 0.765$$

14.4.2 Industry 4.0 Sustainability

This section provides a numerical illustration to demonstrate the proposed methodology for assessing the sustainability index of Industry 4.0 (I4.0)/smart manufacturing from a psychological perspective. The assessment incorporates three critical factors: emerging technologies (ET), technology selection (TS), and site readiness (SR). The distinguishing indicators of sustainability-related issues and performance measurements—existing values and target values (minimum and maximum)—are presented in Tables 14.1 and 14.2 for emerging technologies and technology selection, respectively.

14.4.2.1 For the Emerging Technologies (ET)

The data for emerging technologies (ET), provided in Table 14.1, are normalized and used to calculate the sustainability index for ET. This index serves as a basis for identifying the alignment of ET with the design principles in technology selection (TS). The calculated sustainability index for ET is outlined in Table 14.2, offering a clear mapping between emerging technologies and their role in achieving sustainability within the I4.0 framework.

14.4.2.2 For the Technologies Selection (TS)

Data for emerging technologies is the same as that given in Table 14.1, and the associated result is illustrated in Table 14.2. For mapping the design principles and emerging technologies for Industry 4.0, Equations (10.1)–(10.6), for interoperability (g_1),

TABLE 14.1
Data for the Emerging Technologies

Indicator (f)	Units Performance Measures (%)	Performance Metrics: Existing Value	Target Value: Max	Target Value: Min
f_1	Percentage (%) of DI at a given time t	90	100	80
f_2	Percentage (%) of the IIoT at a given time t	95	100	90
f_3	Percentage (%) of BDA at a given time t	90	95	80
f_4	Percentage (%) of CC/CM at a given time t	85	95	80
f_5	Percentage (%) of cyber security system (CS) at a given time t	95	100	90
f_6	Percentage (%) of VR/AR at a given time t	85	90	80
f_7	Percentage (%) of AM at given time t	95	100	80
f_8	Percentage (%) of CPPS at a given time t	95	100	80

TABLE 14.2
Sustainability Index in Emerging Manufacturing Technologies

Indicator (f)	f_1	f_2	f_3	f_4	f_5	f_6	f_7	f_8
Normalized index	0.50	0.50	0.67	0.25	0.50	0.50	0.75	0.75
$S_{ET}(t)$	0.5525							

virtualization (g_2), decentralization (g_3), real-time capability (g_4), service orientation (g_5), and modularity (g_6), respectively, as follows:

$$g_1 = \frac{f_1 + f_2 + f_3 + f_4 + f_5 + f_7 + f_8}{7}$$

$$= \frac{0.50 + 0.50 + 0.67 + 0.25 + 0.50 + 0.75 + 0.75}{7} = 0.56$$

$$g_2 = \frac{f_1 + f_5}{2} = \frac{0.50 + 0.50}{2} = 0.50$$

$$g_3 = \frac{f_1 + f_7 + f_8}{3} = \frac{0.50 + 0.75 + 0.75}{3} = 0.67$$

$$g_4 = \frac{f_1 + f_2 + f_3 + f_6}{4} = \frac{0.50 + 0.50 + 0.67 + 0.50}{4} = 0.54$$

$$g_5 = \frac{f_1 + f_2 + f_3 + f_4 + f_6}{5} = \frac{0.50 + 0.50 + 0.67 + 0.25 + 0.50}{5} = 0.49$$

$$g_6 = \frac{f_1 + f_4 + f_5 + f_7}{4} = \frac{0.50 + 0.25 + 0.50 + 0.75}{4} = 0.50$$

Equation (10.8) in Chapter 10 is used to assess the sustainability index for emerging technologies selection for Industry 4.0 as follows, estimating the relative weights between interoperability, virtualization, decentralization, real-time capability, service orientation, and modularity by using the Analytical Hierarchy Processes (AHP). The relatives were previously estimated as shown in Figure 10.5 as 21.45%, 17.83%, 16.05%, 20.34%, 11.83%, and 12.50% for interoperability, virtualization, decentralization, real-time capability, service orientation, and modularity, respectively. The sustainability index of technologies selection (TS) at any time t, $s_{TS}(t)$ is evaluated by utilizing Equation (10.8), and the results are 0.55 (55.00%).

$$S_{TS}(t) = 0.2145\ g_1(t) + 0.1783\ g_2(t) + 0.1605\ g_3(t) + 0.2034\ g_4(t) + 0.1183\ g_5(t) + 0.1250\ g_6(t)$$

$$S_{TS}(t) = 0.2145(0.56) + 0.1783(0.50) + 0.1605\ (0.67) + 0.2034\ (0.54) + 0.1183\ (0.49) + 0.1250\ (0.50) = 0.55$$

14.4.2.3 For the Site Readiness (SR)

The illustrated example presented in Section 11.4 (Chapter 11) is used to demonstrate the proposed framework to assess the sustainability index toward I4.0 concerning site readiness (SR). Data related to technical challenges (h_1), risks (h_2), critical success factors (h_3), using of opportunities (h_4), and enablers of implementation (h_5) are presented and illustrated in Tables 11.6–11.10, respectively. The existing current value of $h_1\left(E^{h_1}\right)$ for technical challenges; the existing current value of $h_2(E^{h_2})$; the existing current value of h_3 (E^{h_3}); the existing current value of h_4 (E^{h_4}); and the existing current value of h_5 $\left(E^{h_5}\right)$ are estimated using Equations (11.1)–(11.5) and data in Tables 11.6–11.10 as follows:

$$E^{h_1} = \frac{3\times 3 + 4\times 7 + 5\times 4}{70} = 0.8142$$

$$E^{h_2} = \frac{1\times 2 + 2\times 3 + 3\times 4 + 1\times 5}{50} = 0.60$$

$$E^{h_3} = \frac{2\times 3 + 2\times 4 + 4\times 5}{40} = 0.85$$

$$E^{h_4} = \frac{2\times 4 + 3\times 5}{25} = 0.92$$

$$E^{h_5} = \frac{2\times 2 + 4\times 6 + 5\times 6}{70} = 0.8285$$

TABLE 14.3
Sustainability Index in Site Readiness

Indicator (h)	h_1	h_2	h_3	h_4	h_5
Index of existing value	81.42	60.00	85.00	92.00	82.85
$S_{SR}(t)$	80.25				

The total site readiness issues result appears in Table 14.3. The sustainability index of site readiness (*SR*) at any time *t*, $s_{SR}(t)$ is evaluated by utilizing Equation (12.5), and the results are outlined in Table 14.3 as 80.25%.

The index of sustainability of Industry 4.0/smart manufacturing is assessed, as specified in Equation (12.6) based on the sustainability of emerging manufacturing technologies. Technologies selection and site readiness have equal relative weights as given in the below equation. This implies that the sustainability index in Industry 4.0 is 42%

$$S_{I4.0}(t) = [0.5525]^{\frac{1}{[0.50x0.55+0.50x0.8025]}} = 0.42$$

14.4.3 Manufacturing Sustainability

This section demonstrates the proposed methodology for estimating the sustainable manufacturing index across the entire manufacturing ecosystem. The estimation is based on key components, including machine design, manufacturing processes, and manufacturing systems. These components are analyzed through the identification of sustainability enablers and the evaluation of performance measurements, which include existing values and target values (minimum and maximum). The relevant data and parameters for this assessment are presented in Tables 14.4–14.6. These tables provide detailed information on the sustainability enablers for each component:

Concerning the sustainability of machine design (Table 14.6) and as usual, the foremost imperative issue is normalizing the enablers of sustainability through their indicators. For example, the indicator for tensile quantity (x_1), which is evaluated as the tensile strength at a given time by utilizing two values for the target (greatest as 100 MPa and least as 80 MPa) and one existing value as 95 MPa. At that point, the normalized tensile strength value is assessed as 0.75 as in the following equation, and appears in Table 14.7. The total assessed indicators result in machine design (MD) sustainability, which appears in Table 14.7.

$$x_1(t) = \frac{95-80}{100-80} = 0.75$$

The sustainability index of machine components designs at any time *t*, $s_{MD}(t)$, is evaluated by utilizing Equation (5.1), and the results are outlined in Table 14.7 as 0.4943 (49.43%). The sustainable index of machine design breaks even with 43.43%, and this value appears generally medium, it must be moved forward in all issues/aspects of machine design, either within the introductory design components or

TABLE 14.4
Data for Machine/Product Design

Indicator (x)	Units Performance Measures	Performance Metrics Existing Value	Target Value Max	Min
x_1	Value of tensile strength at a given time t, Mpa	95	100	80
x_2	Value of compression at a given time t, Mpa	100	105	95
x_3	Value of impact at a given time t, Mpa	75	80	55
x_4	Value of hardness at a given time t, Mpa	195	200	190
x_5	Value of toughness at a given time t, Mpa	145	150	140
x_6	Value of torsion at a given time t, Mpa	144	145	142
x_7	Value of shear stress at a given time t, Mpa	148	150	145
x_8	Value of corrosion at a given time t, mm/year	0.008	0.01	0.006
x_9	Value of fatigue at a given time t, Mpa	210	250	190
x_{10}	Value of fracture at a given time t, Mpa	470	500	450
x_{11}	Value of creep at a given time t, mm/year	0.008	0.01	0.006
x_{12}	Value of stiffness at a given time t, Mpa	95	100	90
x_{13}	Value of bending and flexure at a given time t, Mpa	190	200	185
x_{14}	Number of modular modules at a given time t	8	10	6
x_{15}	Possibility to change the head of a machine at a given time t	90	100	80
x_{16}	Number of frames, moving and rotating tables in a machine at any time t	3	4	2
x_{17}	Availability of using the industrial Internet of thing (IIOT) at any time t	90	100	80
x_{18}	Number of sensors at a given time t	12	15	10
x_{19}	Integration of CPS at a given time t	90	95	85
x_{20}	Availability of ECS at a given time t	90	100	80
x_{21}	Availability of DA at a given time t	90	100	80

reconfigurable issues or microsystem technology. Concerning the sustainability of manufacturing processes and systems, the same strategy will be taken after clarification of machine design sustainability for normalized indicators and assessed the sustainability indexes by using Equations (6.1) and (7.1) in previous Chapters 6 and 7, as appeared in Tables 14.8 and 14.9.

The index of sustainable manufacturing is assessed, as specified in Equation (8.1) based on the sustainability of designing machines or products, manufacturing processes analysis and investigation, and facilities design and layout manufacturing systems with equal weights of manufacturing processes and manufacturing systems.

$$S_M(t) = [0.4943]^{\frac{1}{[0.50x0.502+0.50x0.6875]}} = 0.31$$

This assessment indicates that the sustainable manufacturing index rises to 31%, assuming an equal distribution of relative weights between the sustainability of

TABLE 14.5
Data for Manufacturing Processes

Indicator (y)	Units Performance measures	Performance Metrics Existing Value	Target value Max	Min
y_1	The temperature of the oven at any time t, C	1500	2000	1000
y_2	Usage of energy in casting/solidification at any time t, kW	90	100	70
y_3	Utilization (usage) of materials at any time t, %	97	100	95
y_4	Number of products manufactured during the life of the die at any time t	120 K	150 K	100 K
y_5	Usage of energy in bulk deformation, kW	200	250	190
y_6	Availability of designing a new die at any time t for bulk deformation	90	100	85
y_7	Usage of energy in sheet metal deformation at any time t, kW	18	20	15
y_8	Availability of designing a new die at any time t for sheet metal working	90	95	80
y_9	Energy consumption used to machine or cut a workpiece at any time t, kW	20	25	10
y_{10}	Availability of configuration and cutting angles of a cutting tool at any time t	95	100	90
y_{11}	Cutting/machining time before replacing the cutting tool at any time t, minutes	15	25	12
y_{12}	Availability of cooling (coolant) type at any time t	90	95	85

TABLE 14.6
Data for Manufacturing Systems

Indicator (z)	Units Performance Measures	Performance Metrics Existing Value	Target Value Max	Min
z_1	Percentage (%) of using functional (process) manufacturing system in a plant	37	40	30
z_2	Percentage (%) of using flow manufacturing system in a plant	8	10	5
z_3	Percentage (%) of using cellular manufacturing system in a plant	47	50	40
z_4	Percentage (%) of automated systems rather than manual systems	85	90	70

TABLE 14.7
Sustainability Index in Machine Design

Indicator (x)	x_1	x_2	x_3	x_4	x_5	x_6	x_7	x_8	x_9	x_{10}	x_{11}	x_{12}	x_{13}	x_{14}	x_{15}	x_{16}	x_{17}	x_{18}	x_{19}	x_{20}	x_{21}
Normalized index	0.75	0.5	0.67	0.5	0.5	0.66	0.6	0.5	0.5	0.4	0.4	0.5	0.30	0.5	0.2	0.5	0.5	0.4	0.5	0.5	0.5
$S_{MD}(t)$	0.4943																				

TABLE 14.8
Sustainability Index in Manufacturing Processes

Indicator (y)	y_1	y_2	y_3	y_4	y_5	y_6	y_7	y_8	y_9	y_{10}	y_{11}	y_{12}
Normalized index	0.50	0.67	0.60	0.40	0.33	0.33	0.60	0.67	0.67	0.50	0.25	0.50
$S_{MP}(t)$	0.502											

TABLE 14.9
Sustainability Index in Manufacturing Systems

Indicator (z)	z_1	z_2	z_3	z_4
Normalized index	0.70	0.60	0.70	0.75
$S_{MS}(t)$	0.6875			

manufacturing processes and manufacturing systems. However, this result highlights a significant gap of 69% (100%–31%) that must be addressed to achieve full sustainability within the entire manufacturing framework. Bridging this gap will require substantial effort, involving considerable time and financial resources, to elevate the sustainability level across all manufacturing components.

To estimate Sustainability 4.0 (S4.0), either Equations (14.5) or (14.6) can be applied directly using the values calculated above. Based on the calculations, the S4.0 index is estimated to be 48.00%, which is considered a medium level of sustainability. This result highlights the potential for improvement in integrating Industry 4.0 technologies and sustainable manufacturing practices to achieve higher sustainability levels.

$$S4.0(t) = 0.765^{\frac{1}{[0.50 \cdot 0.420 + 0.50 \cdot 0.310]s}} = 0.480 = 48.00\%$$

14.5 CONCLUDING REMARKS

The concept of building Sustainability 4.0 (S4.0) was proposed and introduced through a comprehensive framework rooted in the architecture perspective system of manufacturing, structured around three architectural layers. The core layer integrates academicians, practitioners/industrialists, and governmental agencies, who are identified as the most critical stakeholders in this architectural system of manufacturing. Without the active involvement and collaboration of these stakeholders, the successful implementation and integration of S4.0 cannot be achieved.

The chosen layer, representing sustainability within Industry 4.0, introduces a new dimension to intelligent manufacturing by offering a novel perspective on

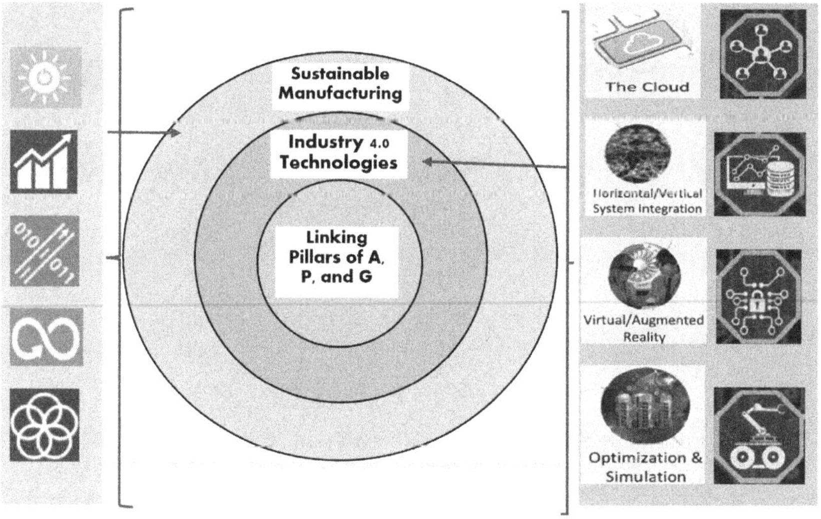

FIGURE 14.7 Outcomes and expectations of achieving sustainability S4.0.

emerging technologies and their applications in advanced or digital manufacturing. This layer highlights the transformative potential of these technologies in achieving sustainability goals.

The final layer, characterized by sustainability in manufacturing, serves as the last essential contribution to the S4.0 architecture system. It ensures that the foundational principles of sustainability are embedded across manufacturing in terms of machine or product design, manufacturing processes, and manufacturing systems. The complete framework for building Sustainability 4.0 is summarized in Figure 14.7. This framework is also mathematically formulated to provide solid evidence of the interrelationships among the layers of S4.0 at any given time.

REFERENCES

Kamble, S. S., and Gunasekaran, A., (2019), "Analyzing the Role of Industry 4.0 Technologies and Circular Economy Practices in Improving Sustainable Performance in Indian Manufacturing Organizations", *Production Planning and Control*, Vol. 34, No. 10, pp. 887–901

Reis, J.S. D.M., Espuny, M., Nunhes, T. V., Sampaio, N.A.D.S., Isaksson, R., Fernando Celso De Campos, F.C., and De Oliveira, O. J., (2021), "Striding Towards Sustainability: A Framework to Overcome Challenges and Explore Opportunities through Industry 4.0", *Sustainability*, Vol. 13, pp. 1–28.

Yadav, G. Kumar, A., Luthra, S., Garza-Reyes, J.A., Kumare, V., and Batista, L., (2020), "A Framework to Achieve Sustainability in Manufacturing Organizations Developing Economies using Industry 4.0 Technologies Enablers", *Computers in Industry*, Vol 122, 103280.

Index

A

Academicians, 140
Additive manufacturing, 81
Adopting strategy for continuous improvement, 105
Adopting vertical integration, 105
Adoption beyond automation, 133
Adoption of wireless-networks systems, 108
Advanced manufacturing solutions, 90
Advanced technologies, 77
Analysis and assessment of sustainable 4.0 selection, 94
Analysis and assessment of sustainable emerging technologies 82
Analysis and assessment of sustainable machine, 41
Analysis and assessment of sustainable site readiness, 108
Applied big data analytical, 108
Architectural layers in Industry 4.0, 11
Architectural layers in sustainable manufacturing, 10
Assembly line, 35
Assessment of manufacturing sustainability, 68

B

Base technologies, 107
Basic technologies, 77
Behavior change, 22
Big data analytics, 80
Building sustainability, 140

C

Cellular manufacturing systems, 35
China, 7
Classification of machine component design, 40
Classification of machining process, 51
Climate action, 3
Cloud computing, 90
Cloud manufacturing, 90
Commitment from upper leadership, 105
Complexity reduction, 135
Components of emerging technologies, 78
Computerized numerical control (CNC), 39
Critical success factors, 104
Cyber-physical production systems (CPPSs), 39, 81

D

Decentralization, 89
Decision making, 23, 29
Design principles of Industry 4.0, 88
Digital and manufacturing technologies, 134
Digital manufacturing (DI), 79
Digitalization, 79

E

Eco-efficiency, 130
Economic pillar, 132
Economic sustainability, 136
Effects of Industry 4.0 technologies on sustainability, 129
Emerging technologies categorization, 90
Emerging technologies sustainability, 119
Enablers of implementation, 107
Enablers of manufacturing processes, 54
Enhance competencies and training of employees, 105
Environmental sustainability, 136

F

Flow line manufacturing systems, 35
Front-end technologies, 107

G

Gender equality, 3
Germany, 7
Governmental agencies, 140

H

Horizontal and vertical integration, 90
Human-centered, 23

I

Industrial organization flexibility level, 105
Industry 2.0 and 3.0, 7
Industry 4.0, 6, 10, 25, 132
Industry 4.0/smart manufacturing streams, 119
Information technology (IT), 39
Innovation and creativity, 23
Integration technologies, 78
Intermediate technologies, 77
Internet of Things (IIoT), 39, 79
Interoperability, 88

J

Job shop, 35, 59

L

Layers of sustainable manufacturing, 36
Lean production and IoT, 135

M

Machine component design, 35
Machine design, 17
Machining processes, 35
Manufacturing processes, 17, 35
Manufacturing processes within an architecture perspective, 52
Manufacturing sustainability layers, 67
Manufacturing systems, 18
Mindset shift, 25–27
Modularity, 89

N

Numerical example, 27

O

Opportunities, 106
Organizational culture, 23, 29

P

Perspective Building of Sustainability 4.0, 142
Physical technologies, 90
Pillar integration and features, 130
Practitioners, 140

R

Real-time capability, 89
Resistance to change, 26
Retrofitting, 35
Risk, 103

S

Service orientation, 89
Shop floor technologies, 78
Site readiness, 20, 100
Site readiness sustainability, 121
Smart manufacturing, 6, 8, 10, 11, 25, 29, 30, 39, 68, 82, 94
Smart storage, 78
Social environmental, 130
Social sustainability, 136
Socio-Economic, 130
Supply chain management, 134
Sustainability 4.0 from a technical perspective, 17
Sustainability 4.0 perspective, 9
Sustainability assessment of Industry 4.0, 121
Sustainability index, 64
Sustainable cities, 3
Sustainable development goals, 3, 5
Sustainable emerging technologies, 77
Sustainable enablers, 41
Sustainable machine design challenges and solutions, 40
Sustainable machine/product design, 39
Sustainable machining processes, 51
Sustainable manufacturing systems, 59
Sustainable manufacturing, 5, 30
Sustainable site readiness/implementation, 100
Sustainable technologies selection, 87
Sweden, 7

T

Technical challenges of implementation, 102
Technologies selection, 20
Technologies selection sustainability, 120
Triple bottom line of manufacturing sustainability, 36

U

United Kingdom, 7
United Nations, 3
United States, 7

V

Virtual environment, 19
Virtual Reality, 20, 79, 82, 88, 92
Virtualization, 88

Printed in the United States
by Baker & Taylor Publisher Services